I0484003

About U.S. Natural Gas Pipelines – Transporting Natural Gas

The U.S. natural gas pipeline network is a highly integrated transmission and distribution grid that can transport natural gas to and from nearly any location in the lower 48 States. The natural gas pipeline grid comprises:

- More than 210 natural gas pipeline systems.
- 300,000 miles of interstate and intrastate transmission pipelines (see mileage table).
- More than 1,400 compressor stations that maintain pressure on the natural gas pipeline network and assure continuous forward movement of supplies (see map).
- More than 11,000 delivery points, 5,000 receipt points, and 1,400 interconnection points that provide for the transfer of natural gas throughout the United States.
- 29 hubs or market centers that provide additional interconnections (see map).
- 394 underground natural gas storage facilities (see map).
- 55 locations where natural gas can be imported/exported via pipelines (see map).
- 5 LNG (liquefied natural gas) import facilities and 100 LNG peaking facilities.

Learn more about the natural gas pipeline network:

Interstate – Pipeline systems that cross one or more States

Intrastate – Pipelines that operate only within State boundaries

Network Design – Basic concepts and parameters

Pipeline Capacity and Usage

Regulatory Authorities

Transportation, Processing, and Gathering

Transportation Corridors – Major interstate routes

Underground Natural Gas Storage – Includes regional breakdowns

Pipeline Development and Expansion

U.S./Canada/Mexico Import & Export Locations

U.S. Natural Gas Pipeline Network

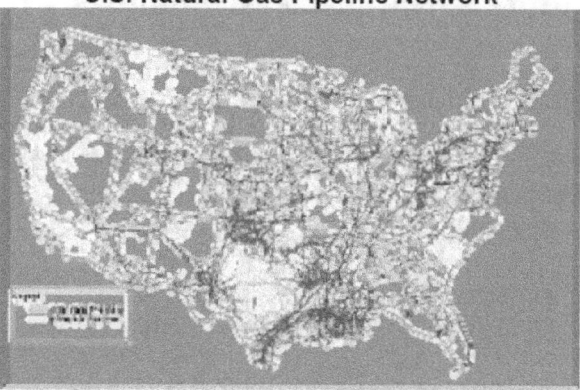

click to enlarge

See Appendix A: Combined 'Natural Gas Transportation' maps

See Appendix B: Tables

Geographic Coverage of Pipeline Companies

United States - links to companies listed A-Z with U.S. map showing regional breakout detail

Northeast - CT, DE, MA, MD, ME, NH, NJ, NY, PA, RI, VA, VT, WV
Midwest - IL, IN, MI, MN, OH, WI
Southeast - AL, FL, GA, KY, MS, NC, SC, TN
Southwest - AR, LA, NM, OK, TX
Central - CO, IA, KS, MO, MT, NE, ND, SD, UT, WY
Western - AZ, CA, ID, NV, OR, WA

About U.S. Natural Gas Pipelines - Transporting Natural Gas

Interstate Natural Gas Pipeline Segment

Two-thirds of the lower 48 States are almost totally dependent upon the interstate pipeline system for their supplies of natural gas.

On the interstate pipeline grid, the long-distance, wide-diameter (20-42 inch), high capacity trunklines carry most of the natural gas that is transported on the national network. In 2005, 85 percent of the 48 trillion cubic feet of gas transported throughout the United States moved through facilities owned by the major interstate pipeline companies. The 30 largest companies own about 77 percent of all interstate natural gas pipeline mileage and about 83 percent of the total capacity (148 billion cubic feet) available within the interstate natural gas pipeline network.

Some of the largest levels of pipeline capacity exist on those natural gas pipeline systems that link the natural gas production areas of the U.S. Southwest with the other regions of the country. Sixteen of the thirty largest U.S. natural gas pipeline systems originate in the Southwest Region, with four additional ones depending heavily upon supplies from the region.

Today, almost every major metropolitan area in the United States is supplied by, or is the final destination of, one or more of the major interstate pipeline companies or their affiliates. For instance,

New York City is a major delivery point on several of the largest pipeline systems, including:
- Texas Eastern Transmission Company
- Transcontinental Gas Company
- Tennessee Gas Pipeline Company, and
- Iroquois Gas Transmission Company.

In the Midwest, Chicago, Illinois, is served by:
- Natural Gas Pipeline Company of America,
- Panhandle Eastern Pipeline Company,
- ANR Pipeline Company,
- Alliance Pipeline Company, and
- Northern Border Pipeline Company.

See regional natural gas pipeline system profiles

Grey States Highly Dependent on Interstate Network

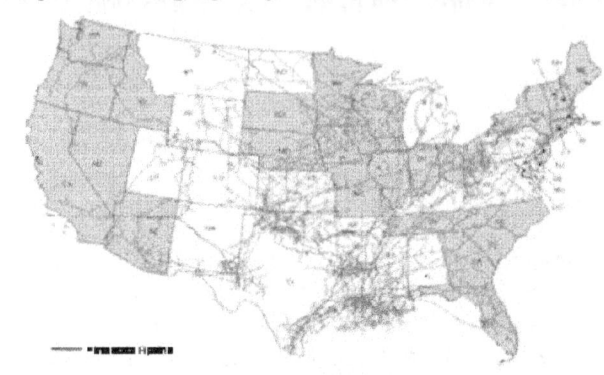

click to enlarge

More information related to interstate pipelines...

Region to Region System Capacity Levels - map
Major Supply Basins & Transportation Corridors - map

Major Interstate Pipeline Companies – table, Appendix B
Pipeline Mileage by State & Region – table, Appendix B

Interstate Pipeline Capacity on a State-to-State Level - spreadsheet

Regional Overviews and Links to Pipeline Companies

Through a series of interconnecting interstate and intrastate pipelines the transportation of natural gas from one location to another within the United States has become a relatively seamless operation. While intrastate pipeline systems often transports natural gas from production areas directly to consumers in local markets, it is the interstate pipeline system's long-distance, high-capacity trunklines that supply most of the major natural gas markets in the United States.

Regional Definitions

click to enlarge

U.S. - all companies listed alphabetically (Appendix B)

Northeast - CT, DE, MA, MD, ME, NH, NJ, NY, PA, RI, VA, VT, WV

Midwest - IL, IN, MI, MN, OH, WI

Southeast - AL, FL, GA, KY, MS, NC, SC, TN

Southwest - AR, LA, NM, OK, TX

Central - CO, IA, KS, MO, MT, NE, ND, SD, UT, WY

Western - AZ, CA, ID, NV, OR, WA

- Of the six geographic regions defined in this analysis, the Southwest Region contains the largest number of individual natural gas pipeline systems (more than 90) and the highest level of pipeline mileage (over 106,000).
- The Central Region produces more gas than it consumes and is a net exporter of natural gas. In recent years, expanding natural gas production in Wyoming, Colorado, New Mexico, and Utah has prompted the construction of several new intra and interstate pipelines in the region and the proposed development of several more over the next several years.
- Although the Midwest Region still receives most of it natural gas supplies from the Southwest Region's producing areas, since the early 1980s new natural gas pipelines from Canada have grown in importance, currently accounting for over one-quarter of the natural gas pipeline capacity entering the region.
- The Northeast Region's natural gas pipeline network has access to supplies from most major domestic gas-producing areas and from Canada. Domestic gas flows into the region from the Southeast into Virginia and West Virginia, and from the Midwest into West Virginia and Pennsylvania. Canadian imports come into the region principally through New York, Maine, and New Hampshire.
- In the Southeast Region most of the twenty-one interstate pipelines serving the region receive most of their supplies from the Gulf of Mexico or from the States of Texas and/or Louisiana.

Slightly more than half the capacity entering the Western Region is on pipeline systems that carry natural gas from the Rocky Mountain area and the Permian and San Juan basins of Texas and New Mexico. These systems enter the region at the New Mexico-Arizona and Nevada-Utah state lines. The rest of the pipeline capacity into the region enters from Wyoming and/or from Canada at the British Columbia-Idaho and Washington State border crossings.

Natural Gas Pipelines in the Northeast Region

Overview | Domestic Gas | Canadian Imports | Regional Pipeline Companies & Links

Northeast Region Natural Gas Pipeline Network

Interstate
Intrastate

Overview

Nineteen interstate natural gas pipeline systems operate within the Northeast Region (Connecticut, Delaware, Massachusetts, Maine, New Hampshire, New Jersey, New York, Pennsylvania, Rhode Island, Virginia, and West Virginia). These interstate pipelines deliver natural gas to several intrastate natural gas pipelines and at least 50 local distribution companies in the region. In addition, they also serve large industrial concerns and, increasingly, natural gas fired electric power generation facilities.

The natural gas pipeline and local distribution companies serving the Northeast have access to supplies from several major domestic natural gas producing areas and from Canada. Domestic natural gas flows into the region from the Southeast into Virginia and West Virginia, and from the Midwest into West Virginia and Pennsylvania. Canadian imports come into the region principally through New York, Maine, and New Hampshire.

Liquefied natural gas (LNG) supplies also enter the region through import terminals located in Massachusetts and Maryland.

Transportation of Domestic Natural Gas Supplies

Almost all of the interstate natural gas pipelines entering or operating within the Northeast Region terminate there as well, including several very large long-distance natural gas pipeline systems that deliver supplies to the region from natural gas producing areas in the U.S. Southwest. The largest of these natural gas pipelines is the Transcontinental Gas Pipeline Company system (8.2 billion cubic feet (Bcf) per day systemwide), which extends from South Texas to the New York City area.

The Tennessee Gas Pipeline Company (6.7 Bcf per day) and Texas Eastern Transmission Company (6.5 Bcf per day) natural gas pipeline systems bring supplies to the Northeast from Texas, Louisiana, and the Gulf of Mexico. The Tennessee Gas Pipeline Company system, unlike the Transcontinental Gas Pipeline Company and Texas Eastern Transmission Company systems, extends its service northward as far as New Hampshire and is a major transporter of natural gas to Connecticut, Massachusetts, and Rhode Island. (Note: The Texas Eastern Transmission Company system was the first natural gas pipeline system to extend from the Southwest to the Northeast Region with the conversion of the "Big-inch" and "Little Big-inch" oil pipelines shortly after World War II.)

The Tennessee Gas Pipeline Company system is also a significant source of supply for the regional Algonquin Gas Transmission Company system, which is the principal interstate natural gas pipeline serving the Boston, Massachusetts area. The Texas Eastern Transmission Company, an affiliate of the Algonquin Gas Transmission Company, is the primary source of supply for that pipeline, delivering approximately 65 percent of Algonquin's requirements at interconnections in New Jersey. The Algonquin Gas Transmission Company system (1,100 miles) has the capability to move 1.1 Bcf per day of its 2.2 Bcf per day system capacity from New Jersey into the New York metropolitan area.

The largest interstate natural gas pipeline system operating in the region is Columbia Gas Transmission Company (8.7 Bcf per day capacity). Columbia has an extensive network of natural gas pipelines that provide service in the region to the States of Maryland, New Jersey, New York, Pennsylvania, Virginia, and West Virginia, but also extends into Ohio in the Midwest and Kentucky and North Carolina in the Southeast Region. Columbia receives Gulf-of-Mexico natural gas at the Kentucky border from its major trunkline transporter, Columbia Gulf Transmission Company, but it also transports Appalachian (regional) production as well.

The Dominion Transmission Company system, although not as extensive as the Columbia Gas Transmission Company system, serves the same States, except for New Jersey. Neither interstate pipeline, however, extends service to the New England States. Both companies are also the major suppliers of some of the largest LDCs in the region, some of which are affiliates.

In addition to the interstate natural gas pipeline companies that bring natural gas into the region, several smaller interstate natural gas pipeline companies operate totally within the region. Among these are systems such as Equitrans Inc. (0.1 Bcf per day), serving West Virginia and western Pennsylvania, that were developed to move local production to regional markets. West Virginia, western Pennsylvania, and southwestern New York were once the region's and the Nation's largest natural gas producing areas and, consequently, have many local gathering, distribution, and storage interconnections. These local facilities also have many interconnections with natural gas pipeline operations in Ohio, which is the reason for the 2.1 Bcf per day of capacity exiting the region to the Midwest.

Some of these smaller interstate and intrastate natural gas pipelines serve niche areas within the region. For instance, Eastern Shore Natural Gas Company (0.1 Bcf per day) is the only natural gas pipeline serving southern Delaware and the lower Delmarva Peninsula. It receives its supplies from Transcontinental Gas Pipeline Company and Columbia Gas Transmission Company at points in southeastern Pennsylvania, with a route that takes it southward through Delaware to Maryland's eastern shore.

Further to the north, Granite State Transmission Company (0.1 Bcf per day) receives natural gas from the Tennessee Gas Pipeline Company and/or the PNGTS/Maritimes and Northeast Pipeline system at the southern New Hampshire/Massachusetts border. From there it transports to customers in New Hampshire and to its NiSource Inc. affiliate, the Northern Utilities Company system -- which has been the historical source of natural gas for the southern portion of the State of Maine. While the KeySpan Energy Delivery Company is the principal provider of natural gas service to New York City and Long Island, New York, its intrastate operations now also include service in Massachusetts and New Hampshire as well.

Importance of Canadian Imports to the Region

During the past 17 years (1990-2005), several new natural gas pipelines were built that substantially increased regional access to Canadian natural gas supplies.

The Iroquois Gas Transmission Company system, completed in 1991, draws almost one Bcf per day off the TransCanada Pipeline Ltd system in Ontario, Canada, a large portion of which is delivered to the New York City metropolitan area. The Empire Pipeline Company system (0.5 Bcf per day), built in 1994, and an intrastate affiliate of National Fuel Gas Supply Corporation, brings in Canadian natural gas at Grand Island, New York (near Niagara Falls) to north central New York State with interconnections to the Dominion Transmission Company and National Fuel Gas Supply Corporation systems. These latter two companies also access Canadian supplies via Tennessee Gas Pipeline, which maintains a 0.9 Bcf per day import point at Niagara Falls, New York. [Map (Appendix A) and Table (Appendix B) of Import/Export Points]

Moreover, in 2000, the Portland Natural Gas Transmission System (PNGTS), which begins at the northern neck of New Hampshire and extends to the coast of Maine, was completed. PNGTS merges with the Maritimes and Northwest Pipeline system at Wells, Maine, where they form a joint 100-mile, 0.6 Bcf per day, natural gas pipeline that extends south through southern New Hampshire and terminates in northern Massachusetts (with an interconnection to the Tennessee Gas Pipeline Company system).

In 2003, PNGTS, originally designed only to import natural gas (into New Hampshire), was reconfigured to provide bidirectional service to its customers. The objective of the reconfiguration was to provide shippers of Canadian Sable Island natural gas, which use the Maritimes & Northeast Pipeline Company system, with an opportunity to redirect some of their natural gas to markets located in Quebec (which previously had access only to western Canadian natural gas supplies).

The U.S. portion of the Maritimes and Northeast Pipeline Company (0.4 Bcf per day) system begins at Calais, Maine, at the Canadian border and extends to Wells, Maine. The full 730-mile Maritimes and Northeast Pipeline system was built to access natural gas production off the eastern coast of Canada (Sable Island) and to transport it to New England. In late 2003, the Maritimes and Northeast Pipeline Company system was extended from Dracut to Boston, Massachusetts, providing an additional 0.2 Bcf per day of service to that metropolitan area.

Several smaller regional natural gas importing pipelines, such as North County Pipeline Company (0.6 Bcf per day) and the St. Lawrence Gas Company (0.1 Bcf per day), both located in upper New York State, and the Vermont Gas Systems Company (0.05 Bcf per day), the only natural gas pipeline in the State of Vermont, depend upon Canadian natural gas imports completely for their natural gas supplies since they do not interconnect with any other U.S. natural gas pipeline.

Principal Natural Gas Pipeline Companies Serving the Northeast Region

Pipeline Name	Principal Supply Source(s)	System Configuration* Primary/Secondary
Interstate & Importing Pipelines		
Algonquin Gas Transmission Co	Interstate System	Trunk/Grid
Columbia Gas Transmission Co	Southwest, Appalachia	Grid
Dominion Cove Point LNG LP	LNG Imports, Interstate System	Trunk
Dominion Transmission Corp	Southwest, Appalachia	Grid/Trunk
Eastern Shore Natural Gas Co	Interstate System	Trunk/Grid
East Tennessee Natural Gas Co	Interstate System	Trunk/Grid
Equitrans Inc	Appalachia, Southwest	Grid
Granite State Gas Transportation Co	Interstate System	Trunk/Grid
Iroquois Gas Transmission Co[1]	Western Canada	Trunk
Maritimes & Northeast Pipeline Co[1]	Eastern Canada	Trunk
National Fuel Gas Supply Corp	Appalachia, Canada	Grid/Trunk
NORA Gas Transmission Co	Interstate System	Trunk
North Country Pipeline Co[1]	Western Canada	Trunk/Grid
Portland Natural Gas Transportation System[1]	Western Canada	Trunk
St. Lawrence Gas Co[1]	Western Canada	Trunk/Grid
Tennessee Gas Pipeline Co[1]	Southwest, Canada	Trunk
Texas Eastern Transmission Corp	Southwest	Trunk
Transcontinental Gas Pipeline Co	Southwest	Trunk
Vermont Gas Systems Inc[1]	Western Canada	Trunk/Grid
*Intrastate Pipelines**		
Empire Gas Pipeline Co (NY)[1]	Canada	Trunk

Dominion Hope Gas Co (WV)	Appalachia, Interstate System	Grid/Trunk
KeySpan Energy Delivery (NY)	Interstate System	Grid
KeySpan Energy Delivery (NH)	Interstate System	Grid
National Fuel Gas Distribution Co (NY)	Interstate System	Grid
NorNew/Norse Pipeline System (NY)	Appalachia, Interstate System	Grid/Trunk
North Penn Gas Co (PA)	Appalachia, Interstate System	Trunk
Northern Utilities Inc (ME)	Interstate System	Trunk/Grid
Penn York Energy Corp (PA)	Appalachia, Interstate System	Trunk
Virginia Natural Gas Co	Interstate System	Trunk/Grid

*System Configuration - natural gas pipeline system design layout. Some systems are a combination of the trunk and grid. Where two are shown, the first represents the predominant system design.

Trunk - systems are large-diameter long-distance trunklines that generally tie supply areas to natural gas market areas.

Grid - systems are usually a network of many interconnections and delivery points that operate in and serve major natural gas market areas.

**Table is not necessarily inclusive of all intrastate natural gas pipelines operating in the region.

[1]Imports and/or exports natural gas between the United States and Canada.

SOURCE: Energy Information Administration, Office of Oil & Gas.

Natural Gas Pipelines in the Midwest Region

Overview | Domestic Gas | Canadian Imports | Regional Pipeline Companies & Links

Overview

Twenty-six interstate and at least seven intrastate natural gas pipeline companies operate within the Midwest Region (Illinois, Indiana, Michigan, Minnesota, Ohio, and Wisconsin). The principal sources of natural gas supply for the region are production areas in the Southwest, although Canadian natural gas pipelines now account for about one-fourth of natural gas pipeline capacity entering the region. Regional natural gas production, principally from Ohio and Michigan, accounts for little more than 8 percent of the gas consumed in the region.

Midwest Region Natural Gas Pipeline Network

—— Interstate
—— Intrastate

Transportation of Domestic Natural Gas Supplies

Traditionally, the principal sources of natural gas for the Midwest Region have been the panhandles of west Texas and Oklahoma, the States of Kansas and Louisiana, and eastern Texas. Approximately 17 billion cubic feet (Bcf) per day of the 28 Bcf per day of peak-day capacity (61 percent) entering the Midwest via the interstate network in 2005 came from production areas in the Southwest Region. The interstate pipeline systems that provide this transportation capacity are some of the largest in the nation.

Two of those pipeline systems, ANR Pipeline Company (ANR) and Natural Gas Pipeline Company of America (NGPLA), operate on corridors that transport supplies from the Texas, Oklahoma, Kansas, and Louisiana production areas. NGPLA provides about 12 percent (3.4 Bcf per day) of the total throughput capacity into the region and terminates in the Chicago, Illinois, area. ANR can transport 2.0 Bcf per day into the region and operates in all Midwest States except Minnesota, terminating in Michigan and Indiana.

Three systems, Northern Natural Gas Company, Panhandle Eastern Pipeline Company, and Centerpoint Mississippi River Transmission Company transport gas to the Midwest from the Texas/Oklahoma/Kansas production area, while four others, Texas Gas Transmission Company, Trunkline Gas Company, Texas Eastern Pipeline Company, and Tennessee Gas Pipeline Company systems begin in Louisiana and east Texas and proceed directly north into the Midwest Region. However, most of the capacity on the latter two systems is intended for markets in the Northeast, with few or no deliveries within the Midwest Region itself.

The two most recent additions to the regional network are the Horizon Pipeline Company (0.4 Bcf per day) and the Guardian Pipeline Company (0.8 Bcf per day) systems, both completed in 2002. Each receives natural gas supplies in the Chicago area from the interstate natural gas pipeline system for delivery to expanding markets in northern Illinois and the greater Milwaukee, Wisconsin, metropolitan area. The Crossroads Pipeline Company (0.3 Bcf per day), an affiliate of Columbia Gas Transmission Company, provides natural gas transportation for shippers seeking a route between interstate natural gas pipelines serving western Indiana to interconnections in central Ohio to the Columbia Gas Transmission Company system. For the most part, natural gas customers (and LDCs) in Indiana are dependent upon interstate pipelines that traverse the State. Only the Texas Gas Transmission Company system terminates in the State, where it directs about 30 percent of its total system capacity.

In the eastern half of the region, such as in Ohio, the interstate Dominion Transmission Company and Columbia Gas Transmission Company systems predominate. Each system provides extensive service to affiliated local distribution companies (LDCs) (NiSource and Dominion), and in the case of Columbia, access to 15 underground natural gas storage sites located in Ohio. Both systems are essentially extensions of natural gas pipeline operations that developed historically in Pennsylvania and West Virginia. In fact, a portion of the natural gas delivered in Ohio from these systems originates with natural gas production in Appalachia.

Importance of Canadian Natural Gas Imports to the Region

Seven interstate natural gas pipeline companies transport Canadian natural gas into or out of the Midwest (see Table below) with a combined capacity of approximately 8.4 Bcf per day, or about 30 percent of the total interstate capacity entering the region. A decade ago, import capacity represented only about 10 percent of the capacity into the region.

The largest natural gas importing pipeline in the region is the Great Lakes Gas Transmission Company (2.9 Bcf per day) system, which links to the TransCanada Pipeline Ltd system at the Manitoba/Minnesota border and proceeds through the northern portion of Minnesota, Wisconsin, and Michigan and southward through Michigan to the Michigan/Ontario border. [Map and Table of Import/Export Points]

However, a large portion (about 85 percent) of the natural gas transported on the Great Lakes Transmission Company system is delivered back into Canada for consumption in Ontario and eastern Canada. In contrast, Viking Gas Transmission Company (0.5 Bcf per day) receives Canadian natural gas at the same Manitoba/Minnesota border point as Great Lakes Transmission Company (Noyes, Minnesota), but its volumes are delivered and consumed entirely within the United States with deliveries to eastern North Dakota, Minnesota, and central Wisconsin.

Natural gas transportation on the Northern Border Pipeline Company system (2.5 Bcf per day) reaches the Midwest Region by way of the Central Region (from the Saskatchewan/Montana border, through North Dakota, South Dakota, Minnesota, and Iowa, -- then into Illinois and finally western Indiana). The Northern Border Pipeline Company system physically reached the Midwest for the first time in 1998 with completion of a 200-mile, 0.7 Bcf per day extension from Iowa (Central Region) to the vicinity of Chicago, Illinois. In the Chicago area, the Northern Border Pipeline system interconnects with several other interstate pipelines and with several LDCs, including NICOR and Peoples Gas & Light Company.

In 2001, Northern Border Pipeline Company system also interconnected with the new Vector Pipeline Company at the Chicago hub, permitting its shippers of western Canadian (Alberta) natural gas an alternative route to reach delivery points in Ontario, Canada. Subsequently, the Northern Border Pipeline Company system was also extended (34 miles, 0.5 Bcf per day) to just east of the Indiana/Illinois border where it provided its shippers access to the Indiana/Ohio market with direct interconnections to customers such as the Northern Indiana Public Service Company (NIPSCO). The new extension also included local interconnections with other interstate natural gas pipeline systems in Indiana.

A sizable portion of the natural gas transported on the Northern Border Pipeline Company system still reaches the Midwest indirectly through interconnections with the Northern Natural Gas Company and Natural Gas Pipeline Company of America interstate systems in Iowa. The Northern Natural Gas Company, which also provides shippers with natural gas transportation services to Midwest markets from the Southwest Region's production fields, operates an extensive network of pipelines in southern Minnesota, Wisconsin, and in northwestern Illinois.

For the most part, the natural gas that flows on the Centra Pipelines Minnesota Company system, the smallest (0.06 Bcf per day) of the natural gas importing pipelines, does not remain in the United States. Instead, this natural gas pipeline system transports gas to Canadian customers located in southwestern Ontario who do not have access to other Canadian natural gas pipeline sources. The Centra Pipelines Minnesota Company receives its Canadian natural gas supplies at the Manitoba/Minnesota border. It then crosses the northern section of Minnesota on an eastward route until it re-enters Canada at the Minnesota/Ontario border.

The newest large capacity pipeline to import Canadian natural gas into the region is the Alliance Pipeline Company system (2.1 Bcf per day). Completed in late 2000, the U.S. portion of the Alliance system extends from the Saskatchewan/North Dakota border southeast through Minnesota and Iowa, and terminates in the vicinity of Joliet, Illinois, at the Aux Sable natural gas processing plant.

The "wet" gas that is processed at the Aux Sable processing plant is delivered "dry" (pipeline quality natural gas) at its tailgate to several major interstate natural gas pipelines, including ANR Pipeline Company, Natural Gas Pipeline Company of America, Midwestern Gas Transmission Company, and Vector Pipeline Company, for shipment to customers in Illinois, Indiana, Ohio, Michigan, and Ontario, Canada. In addition, several intrastate natural gas pipeline companies, and LDCs, including Peoples Gas & Light Company and Northern Illinois Gas Co (NICOR), receive natural gas at the Aux Sable tailgate.

Principal Natural Gas Pipeline Companies Serving the Midwest Region

Pipeline Name	Principal Supply Source(s)	System Configuration* Primary/Secondary
Interstate & Importing Pipelines		
Alliance Pipeline Co[1]	Canada	Trunk
ANR Pipeline Co	Louisiana, Kansas, Texas	Trunk/Grid
ANR Storage Co	Michigan, Interstate System	Trunk
Centerpoint Mississippi River Trans Corp	Arkansas, Oklahoma	Trunk
Centra Pipelines Minnesota Co[1]	Canada	Trunk
Columbia Gas Transmission Co	Louisiana, Appalachia	Grid/Trunk
Crossroads Pipeline Co	Interstate System	Trunk
Dominion Gas Transmission Co	Louisiana, Appalachia	Grid/Trunk
Great Lakes Gas Transmission Ltd[1]	Canada	Trunk
Guardian Pipeline Co	Interstate System	Trunk
Horizon Pipeline Co	Interstate System	Trunk
KO Gas Transmission Co	Interstate System	Trunk
MichCon Gas Co	Michigan, Interstate System	Trunk
Midwestern Gas Transmission Co	Interstate System	Trunk
Missouri Interstate Gas LLC	Interstate System	Grid/Trunk
Natural Gas PL Co of America	Kansas, Oklahoma, Louisiana, Texas	Trunk
NGO Gas Transmission Co	Interstate System	Trunk
Northern Border Pipeline Co[1]	Canada	Trunk
Northern Natural Gas Co	Kansas, Oklahoma, Texas	Trunk/Grid
Panhandle Eastern PL Co[1]	Kansas, Oklahoma, Texas	Trunk
Tennessee Gas Pipeline Co	Louisiana, Texas	Trunk
Texas Eastern Transmission Co	Louisiana, Texas	Trunk
Texas Gas Transmission Co	Louisiana, Texas	Trunk
Trunkline Gas Co	Louisiana, Texas	Trunk
Viking Gas Transmission Co[1]	Canada	Trunk
Vector Pipeline Co[1]	Interstate System	Trunk
*Intrastate Pipelines**		

Cardinal Pipeline Co (IN)	Interstate System	Trunk
Consumers Gas Co (MI)	Interstate System	Grid/Trunk
Dominion East Ohio (OH)	Interstate System	Grid
Northern Indiana Public Service Co	Interstate System	Grid/Trunk
NorthCoast Gas Transmission Co (OH)	Interstate System	Trunk
Northern Illinois Gas Co (NICOR) (IL)	Interstate System	Trunk/Grid
Saginaw Bay Pipeline (MI)	Interstate System	Trunk/Grid

*System Configuration - natural gas pipeline system design layout. Some systems are a combination of the trunk and grid. Where two are shown, the first represents the predominant system design.
 Trunk - systems are large-diameter long-distance trunklines that generally tie supply areas to natural gas market areas.
 Grid - systems are usually a network of many interconnections and delivery points that operate in and serve natural gas major market areas.
**Table is not necessarily inclusive of all intrastate natural gas pipelines operating in the region.
[1]Imports and/or exports natural gas between the United States and Canada.
SOURCE: Energy Information Administration, Office of Oil & Gas.

Natural Gas Pipelines in the Southeast Region

Overview | Transportation to Atlantic & Gulf States | Gulf of Mexico Transportation Corridor | Transportation to the Northern Tier | Regional Pipeline Companies & Links

Overview

Twenty-two interstate, and at least nine intrastate, natural gas pipeline companies operate within the Southeast Region (Alabama, Florida, Georgia, Kentucky, Mississippi, North Carolina, South Carolina, and Tennessee). Fifteen of the twenty-one interstate natural gas pipelines originate in the Southwest Region and receive most of their supplies from the Gulf of Mexico or from the States of Texas and/or Louisiana.

Southeast Region Natural Gas Pipeline Network

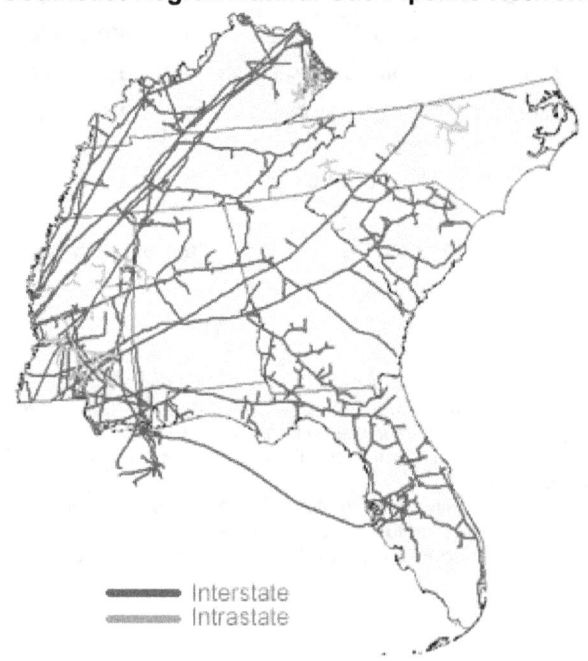

Interstate
Intrastate

Of those fifteen, nine natural gas pipelines (see Table below) actually flow all or a substantial portion, of their deliveries beyond the region, to points within the Northeast or Midwest regions. The remaining natural gas pipelines companies, with the exception of Columbia Gas Transmission Company which is primarily a Northeast Regional pipeline that also supplies a small territory in northern North Carolina, serve the needs of natural gas shippers and customers within the region itself.

Transportation of Natural Gas to Atlantic & Gulf States

By capacity level, the largest natural gas transporters in the region are the Transcontinental Gas Pipeline Company system, Tennessee Gas Pipeline Company, Texas Eastern Transmission Company, and Texas Gas Pipeline Company. The Transcontinental Gas Pipeline Company is the second largest natural gas pipeline in the United States with an overall capability to transport up to 8.2 billion cubic feet (Bcf) per day of natural gas. Although a major portion of its deliveries are directed into the Northeast Region, the Transcontinental Gas Pipeline Company is still the single source of natural gas supplies to many customers in the region, especially in North and South Carolina. This system begins in southeast Texas and extends northeastward through Louisiana, with regional service in Mississippi, Alabama, and the Atlantic coastal States of Georgia, and North and South Carolina.

The largest natural gas pipeline system serving primarily the Southeast is the Southern Natural Gas Company (3.4 Bcf per day). Its system transports natural gas supplies from Southwestern production areas to customers in Mississippi, Alabama, Georgia, and South Carolina and to a small section of southern Tennessee. With interconnections provided by its subsidiary, South Georgia Natural Gas Company (0.1 Bcf per day), the Southern Natural Gas Company also provides transportation service to parts of northern Florida.

Southern Natural Gas Company is the major supplier of natural gas to the Atlanta Gas Light Company (Georgia) and the South Carolina Gas Company, two of the largest local distribution companies (LDCs) in the region. The Gulf South Pipeline Company also is a major transporter of natural gas in the southern portion of the region, providing transportation service to and within Mississippi, Alabama, and the western panhandle.

In 2002, the Elba Island, Georgia, LNG import facility reopened after being closed since 1980. To provide transportation services to a key customer of the facility, the SCG Pipeline Company system, now part of the Carolina Gas Transmission Corporation, built a 0.2 Bcf per day, 18-mile natural gas pipeline in 2003 between Elba Island and a new 875 MW natural gas fired power plant located in southeastern South Carolina. This natural gas pipeline, in addition to a restored twin-pipeline system between the Elba Island facility and Southern Natural Gas Company's system in Georgia, now can deliver up to 0.8 Bcf per day into the Carolina Gas Transmission Corporation system (which was reclassified as an interstate system in 2004).

Gulf of Mexico Natural Gas Transportation Corridor

The 1.1 Bcf per day Gulfstream Natural Gas Pipeline Company system became operational in June 2002. It has the capability to transport natural gas supplies from the Mobile Bay area of Alabama across the Gulf of Mexico to points in west central Florida. Since 2002, the portion of this natural gas pipeline system within the State of Florida has been extended to the east coast of the State as well, although some initial delays in power plant development there slowed its final phases.

The completion of the Gulfstream Natural Gas Pipeline Company system meant that Florida Gas Transmission Pipeline Company was no longer the only source of natural gas available to the State's natural gas shippers and customers. Yet, given the increasing demand for natural gas in Florida, especially from new power generation plants, the Florida Gas Transmission Pipeline Company continues to expand its own system. Between 2000 and 2005, it installed 0.7 Bcf per day of new capacity and will add another 0.2 Bcf per day in 2007 with completion of its Phase VII expansion. The entire system extends from southeastern Texas, through southern Louisiana, Mississippi, Alabama, and into and throughout Florida.

A significant portion of the region's natural gas pipeline capacity, as well as a primary source of natural gas supplies for the Gulfstream Natural Gas Pipeline Company system, comes from the Gulf of Mexico on three natural gas pipeline systems: the Dauphin Island Gathering System (1.2 Bcf per day), the Destin Pipeline L. P. (1.2 Bcf per day) and the Chandeleur Pipeline Company (0.33 Bcf per day). The latter two systems transport natural gas from Gulf of Mexico production areas to onshore delivery points at Pascagoula, Mississippi, while the Dauphin Island Gathering System directs its flow to Alabama and interconnections with at least six major interstate systems, including the Gulfstream Natural Gas Pipeline Company system.

Natural Gas Transportation Services within the Northern Tier

The northern tier of the Southeast Region is served primarily by regional interstate natural gas pipelines such as the East Tennessee Natural Gas Company and the Enbridge Pipelines (AlaTenn) Inc. (formerly Alabama-Tennessee Gas Pipeline Company) systems, who, in turn, receive their supplies mainly from those interstate natural gas pipelines that traverse the region through Tennessee and Kentucky. These interconnecting natural gas supply pipelines include the ANR Pipeline Company, Columbia Gulf Transmission Company, Midwestern Transmission Company, Tennessee Gas Pipeline Company, Texas Eastern Transmission Company, Texas Gas Transmission Company, and the Trunkline Gas Company systems.

These natural gas pipelines, with the exception of interconnections with regional natural gas pipelines and deliveries to some large industrial facilities and natural gas fired electric power generation facilities within the region, reserve the major share of system deliverability for delivery points in the north and east of the region. Columbia Gulf Transmission Company, for example, delivers more than 90 percent of its transported volumes to its affiliate, Columbia Gas Transmission Company, at the Kentucky/West Virginia border; however, over the past several years, it has also constructed laterals to supply new natural gas fired power plants built along its route.

Of the several intrastate pipelines operating within the region, three are primarily transporters of local natural gas production to local distribution networks or provide interconnections with other natural gas pipelines. For instance, Tengasco Pipeline Company links production from the developing Swan Creek field in Hawkins county, Tennessee with local municipalities, and can extend transportation services to shippers with destinations in Virginia via interconnections with the East Tennessee Natural Gas Company system.

In Mississippi, the Atmos Energy Gas Company (formerly Mississippi Valley Gas Company) not only provides transportation from the growing conventional natural gas fields in the area but also from the developing coal-bed methane production sources in the State as well. Similarly, the Enterprise Intrastate-Alabama Pipeline Company owns/operates about 450 miles of natural gas gathering and transmission pipelines in the coal-bed methane rich Black Warrior Basin in western Alabama. The system gathers supplies for delivery to the Southern Natural Gas Company and Tennessee Gas Pipeline Company systems, as well as to regional local distribution companies and municipalities.

Principal Natural Gas Pipeline Companies Serving the Southeast Region

Pipeline Name	Principal Supply Source(s)	System Configuration* Primary/Secondary
Interstate Pipelines		
ANR Pipeline Co[1]	Louisiana, Texas	Trunk/Grid
B-R (USG) Pipeline Co	Interstate System	Trunk
Carolina Gas Transmission Corp	Interstate System	Trunk/Grid
Chandeleur Pipeline Co	Gulf of Mexico	Trunk
Columbia Gas Transmission Corp [1]	Appalachia, Interstate System	Grid/Trunk
Columbia Gulf Transmission Co[1]	Gulf of Mexico, Louisiana, Texas	Trunk
Dauphin Island Gathering System	Gulf of Mexico	Trunk
Destin Pipeline LP	Gulf of Mexico	Trunk
East Tennessee Natural Gas Co	Interstate System	Grid/Trunk
Enbridge Pipelines (AlaTenn)	Interstate System	Trunk
Enbridge Pipelines (MidLa)	Louisiana, Mississippi	Trunk
Florida Gas Transmission Co	Louisiana, Texas, Mississippi	Trunk
Gulf South Pipeline Co	Arkansas, Louisiana, Texas	Trunk/Grid
Gulfstream Natural Gas Pipeline Co	Interstate System, Gulf of Mexico	Trunk
Midwestern Gas Transmission Co[1]	Interstate System	Trunk
South Georgia Natural Gas Co	Interstate System	Trunk/Grid
Southern Natural Gas Co	Louisiana, Texas	Trunk/Grid
Tennessee Gas Pipeline Co[1]	Gulf of Mexico, Louisiana, Texas	Trunk
Texas Eastern Transmission Corp[1]	Gulf of Mexico, Louisiana, Texas	Trunk
Transcontinental Gas Pipeline Co[1]	Gulf of Mexico, Louisiana, Texas	Trunk
Texas Gas Transmission Co[1]	Gulf of Mexico, Louisiana	Trunk
Trunkline Gas Co[1]	Louisiana, Texas	Trunk
Intrastate Pipelines **		
Cardinal Pipeline Co (NC)	Interstate System	Trunk
Central Kentucky Transmission (KY)	Interstate System	Trunk
Enterprise Intrastate-Alabama (AL)	Alabama Production	Trunk/Grid
Enbridge Pipelines (Alabama Intra)	Alabama Production	Trunk/Grid
Atmos Energy Gas Co (MS)	Mississippi	Grid
Pub Svc Co of North Carolina (NC)	Interstate System	Trunk/Grid
Sandhill Pipeline Co (NC)	Interstate System	Trunk
Tengasco Pipeline Co (TN)	Tennessee Production	Trunk

*System Configuration - natural gas pipeline system design layout. Some systems are a combination of the trunk and grid. Where two are shown, the first represents the predominant system design.
Trunk - systems are large-diameter long-distance trunklines that generally tie supply areas to natural gas market areas.

Grid - systems are usually a network of many interconnections and delivery points that operate in and serve major natural gas market areas.

**Table is not necessarily inclusive of all intrastate natural gas pipelines operating in the region.

[1]Natural gas pipeline system that transports all or a substantial portion of its deliveries to the Northeast or Midwest Regions.

SOURCE: Energy Information Administration, Office of Oil & Gas.

Natural Gas Pipelines in the Southwest Region

Overview | Export Transportation | Intrastate | Connection to Gulf of Mexico | Regional Pipeline Companies & Links

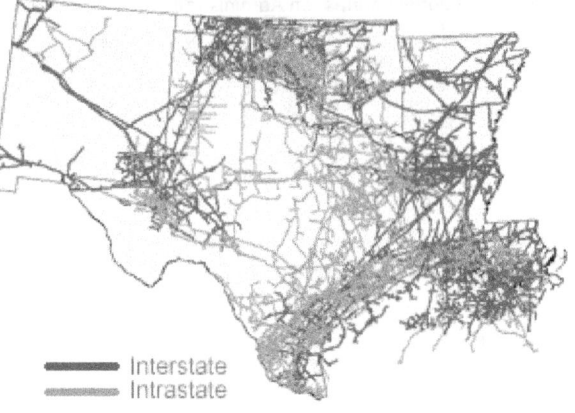

Southwest Region Natural Gas Pipeline Network

━━━ Interstate
━━━ Intrastate

Overview

Most of the major onshore interstate natural gas pipeline companies (see Table below) operating in the Southwest Region (Arkansas, Louisiana, New Mexico, Oklahoma, and Texas) are primarily exporters of the region's natural gas production to other parts of the country and Mexico, while an extensive Gulf of Mexico and intrastate natural gas pipeline network is the main conduit for deliveries within the region. More than 55,000 miles of natural gas pipeline on more than 64 intrastate natural gas pipeline systems (including offshore-to-onshore and offshore Gulf of Mexico pipelines) deliver natural gas to the region's local natural gas distribution companies and municipalities and to the many large industrial and electric power facilities located in the region.

Natural Gas Export Transportation Services Predominate in the Region

Some of the largest and oldest natural gas pipeline systems in the United States originate in the Southwest Region. During the 1930s, the first long-distance natural gas trunklines to serve the Midwest Region from the prolific Hugoton Basin located in the Texas/Oklahoma Panhandle were built. The initial section of the Panhandle Eastern Pipeline Company system reached as far east as central Indiana, although it did not yet extend into Michigan and provide exports of natural gas to Canada as it does today. [Map and Table of Import/Export Points] The initial version of the Natural Gas Pipeline Company of America system extended natural gas service to as far east as Chicago, Illinois, while that of the Northern Natural Gas Company pipeline extended north only into central Iowa and central Minnesota. Today these three systems have the capability to deliver a combined capacity of up to 4.0 billion cubic feet (Bcf) per day to delivery points in the Midwest and Central Regions of the country.

Interstate natural gas pipeline service extended from the Southwest Region to the Northeast Region during the 1940s. The Texas Eastern Transmission Company system began its existence as two oil pipeline systems built in 1943, which were referred to as the "Big Inch" and "Little Big Inch" pipelines. These two systems, converted to natural gas in 1947, extended from Texas to the Midwest and eastward to New York/New Jersey.

The Tennessee Gas Pipeline Company system, which follows a more southerly route than the Texas Eastern system, was also built in the 1940s. It too was developed to supply natural gas to the Northeast, where production fields located in Appalachia (West Virginia, Pennsylvania, New York) were unable to meet the growing demand for natural gas in the region. Subsequently, the Transcontinental Gas Pipeline Company and Columbia Gulf Transmission Company systems also were built to meet the need for long-distance gas transportation between the Southwest and Northeast Regions.

Several interstate natural gas pipeline systems originating in the Southwest also serve the Southeast Region's coastal States. These include Florida Gas Transmission Company, Gulf South Pipeline Company, Southern Natural Gas Company (SONAT), and Transcontinental Gas Pipeline Company, with the latter extending its reach to the Northeast Region as well.

The highly integrated natural gas pipeline network located on the Texas gulf coast and southern Louisiana is a mixture of intrastate natural gas pipelines with interconnections to the interstate network that also serve a large domestic natural gas industrial base and a growing natural gas fired electric generation market. The area has a web of natural gas pipeline linkages, a number of which have developed around several local natural gas market centers/hubs, the largest being the Carthage, Henry, and Egan hubs located in eastern Texas and southwestern Louisiana. Exiting this area of Louisiana, natural gas flows are basically directed either to the east (and to markets along the Atlantic and Gulf coasts and the Northeast Region), or northward through Mississippi, Tennessee, Kentucky, and eventually Illinois and other areas of the Midwest Region.

In the latter case, several of the interstate natural gas pipeline companies using this south-to-north transportation corridor (ANR Pipeline Company, Midwestern Gas Transmission Company, Natural Gas Pipeline Company of America, Trunkline Gas Company, and Texas Gas Transmission Company) provide shippers and gas traders with transportation links between the two largest natural gas market centers in North America, the Henry Hub in Louisiana and the Chicago Hub in Illinois. Natural gas also flows into this Henry-Chicago corridor from the East Texas/North Louisiana production basins and the Arkcoma/Anadarko basin of Oklahoma via natural gas pipelines such as the CenterPoint Gas Transmission Company and Ozark Transmission Company systems.

Although large volumes of natural gas leave the Southwest Region for other regional markets, significant volumes still are consumed in the region. In 2005, the Southwest Region consumed more natural gas than any other region of the country, double that of the next largest consuming region, the Midwest. Indeed, about one-third of the Nation's gas is consumed in the Southwest, in part, owing to the region's high industrial demand for natural gas that has developed over the years because of the ease of access to the many prolific natural gas production fields located in the region.

Only a few pipelines bring natural gas into the Southwest Region. About 4.6 Bcf per day of capacity enters the region compared with the 41 Bcf per day that exits it. One of the few is the Colorado Interstate Gas Company, which is expanding its development of coal-bed methane production in the Raton Basin of southeastern Colorado. Over the past several years (2000-2005) the company has increased its natural gas pipeline capacity southeast into Oklahoma and the Texas panhandle to about 0.6 Bcf per day, primarily to serve newly installed natural gas fired electric generation power plants in the area. The Southern Star Central Gas Pipeline Company (formerly Williams Gas Pipeline Central) system, with operations primarily in Kansas, also directs about 0.3 Bcf per day of natural gas pipeline capacity into the north central portion of Oklahoma.

Another 2.2 Bcf per day of natural gas pipeline capacity entering the region represents capacity from the San Juan Basin and Piceance Basin of Colorado (Central Region) into New Mexico, most of which is redirected toward markets in Arizona and California through the El Paso Natural Gas Company, Transwestern Pipeline Company, and TransColorado Gas Transmission Company systems. The remaining entering capacity, for the most part, represents natural gas import capacity from Mexico (at bi-directional border crossings on the Texas Eastern Transmission Pipeline Company (0.4 Bcf per day), Tennessee Gas Pipeline Company (0.3 Bcf per day), and El Paso Natural Gas Pipeline Company (0.2 Bcf per day) systems), but also includes bi-directional operational flows among the natural gas pipelines that operate between the producing areas of Louisiana and Mississippi basins, such as the Gulf South Pipeline Company and Enbridge Pipelines (Mid-LA) LP systems.

The Region's Intrastate Natural Gas Pipeline Network Is Crucial

The extensive intrastate natural gas pipeline network in the Southwest provides much of the transportation services between the region's producing basins and the interstate network of exporting natural gas pipelines. In many instances they do so with interconnections at one or more of the 18 natural gas market centers/hubs located throughout the Southwest Region. Local natural gas production not only enters the interstate network from these intrastate natural gas pipelines, but several of the major interstate natural gas pipeline systems that terminate in the Northeast, Midwest, and Southeast Regions, originate in Southwest production basins.

The market center/hub links provide transportation flexibility to natural gas shippers wishing to reach alternative markets. For example, the Guadalupe Pipeline Company (formerly TECO), and Oasis Pipeline Company systems provide transportation between west (WAHA hubs) and southeast Texas (Katy Hub), while the Atmos Pipeline--Texas Company (formerly TXU Lone Star), provides a similar service between the same WAHA hubs but to interconnections in northeast Texas (Carthage Hub). Shippers using the Enterprise Texas Intrastate Company system, the portion that was formerly Valero Transmission Company, can deliver to either the Katy or Carthage hubs from the WAHA area.

In West Texas, in the Permian Basin (WAHA) area near the New Mexico border, intrastate natural gas systems (such as the Southern Union Intrastate Pipelines and the Oasis Pipeline Company) interconnect with the interstate network via the El Paso Natural Gas Company, Transwestern Pipeline Company, Natural Gas Pipeline Company of America and the Northern Natural Gas Company pipeline systems. The first two companies provide shippers access to Arizona and California markets, while the latter two provide transportation to Midwest markets. The El Paso Natural Gas Company and Transwestern Pipeline Company systems are also the principal sources of natural gas for the Pubic Service Company of New Mexico, the primary intrastate natural gas pipeline in New Mexico.

In the eastern half of Texas, large intrastate natural gas pipelines such as Enbridge Pipelines (North Texas), Kinder Morgan Texas Pipeline, and Houston Pipeline Companies, provide area producers the opportunity to deliver their natural gas into the interstate network or to customers in major local natural gas consuming areas, such as the metropolitan Houston/Galveston area with its large petrochemical industrial complex. Several natural gas pipelines, including Kinder Morgan Border Pipeline Company and Norteno Pipeline Company, also provide shippers access to natural gas export points to Mexico.

In recent years, the expanding development of natural gas resources in the Barnett Shale formation in the Fort Worth Basin of northeast Texas has supported the installation of several new intrastate natural gas pipelines in the area and the expansion of others, in addition to new or expanded gathering systems. Developers such as Energy Transfer Partners LP and Crosstex Energy Services Company are among the natural gas pipeline companies and joint ventures that have improved access to the interstate and intrastate natural gas transportation networks for producers operating in the growing Forth Worth Basin. Moreover, anticipated production growth in the Basin has generated plans for additional natural gas pipeline expansions in the area over the nexty five years.

Gulf of Mexico Connections to Onshore Texas and Louisiana

Twelve of the largest interstate natural gas systems in the Southwest Region also have operations in the Gulf of Mexico (see above Table), where they either transport natural gas from production platforms or they have interconnections to natural gas gathering systems which are networked to one or more producing platforms or leases. The latter case is characteristic of many of the interconnections that have developed over the past decade, where natural gas exploration and development has shifted into the deepwater portions of the Gulf.

Consequently, a number of new high-capacity natural gas pipelines have been built to bring offshore production directly onshore from associated natural gas gathering operations. For example, the Discovery Gas Transmission Company (0.6 Bcf per day), Destin Pipeline Company (1.0 Bcf per day), and Nautilus Pipeline System (0.6 Bcf per day) each extend from far offshore to onshore interconnections with the existing interstate natural gas pipeline network. In many instances, natural gas production is fed into these offshore/onshore pipelines from large offshore gathering systems that were installed over the past 7 years (2000-2006) to support newer production fields developed in the deepwater. Among these are the East Breaks Gathering System, the Okeanos Deepwater System, and the Cleopatra Gathering System.

Principal Natural Gas Pipeline Companies Serving the Southwest Region

Pipeline Name	Principal Supply Source(s)	System Configuration* Primary/Secondary
Interstate & Importing Pipelines		
ANR Pipeline Co[1]	Louisiana, Kansas, Texas, Gulf of Mexico	Trunk/Grid
Centerpoint Energy Pipeline Co	Kansas, Oklahoma, Texas	Trunk/Grid
Centerpoint Mississippi River Trans	Arkansas, Oklahoma	Trunk

Co

Colorado Interstate Gas Co	Colorado, Oklahoma, Texas	Trunk/Grid
Columbia Gulf Transmission Co[1]	Louisiana, Gulf of Mexico	Trunk
El Paso Natural Gas Co	San Juan (CO,NM) & Permian Basin (TX)	Trunk
Enbridge Pipelines (MidLa)	Louisiana, Mississippi	Trunk
Florida Gas Transmission Co[1]	Louisiana, Texas, Gulf of Mexico	Trunk/Grid
Gulf South Pipeline Co[1]	Louisiana, Mississippi, Texas, Gulf of Mexico	Trunk/Grid
KM Interstate Gas Transmission Co	Kansas, Oklahoma, Wyoming	Trunk/Grid
Natural Gas PL Co of America[1]	Oklahoma, Louisiana, Texas, Gulf of Mexico	Trunk
Northern Natural Gas Co[1]	Kansas, Oklahoma, Texas, Gulf of Mexico	Trunk/Grid
OkTex Pipeline Co	Oklahoma, Northwest Texas	Trunk/Grid
Ozark Gas Transmission Co	Arkansas, Oklahoma	Trunk
Panhandle Eastern PL Co	Kansas, Oklahoma, Texas	Trunk
Sabine Pipeline Co[1]	Texas, Louisiana, Gulf of Mexico	Trunk
Southern Natural Gas Co[1]	Arkansas, Louisiana, Texas, Gulf of Mexico	Trunk/Grid
Southern Star Central Gas Pipeline Co	Kansas, Oklahoma, Texas	Trunk/Grid
Southern Trails Pipeline Co	New Mexico (San Juan Basin)	Trunk
Tennessee Gas Pipeline Co[1,2]	Louisiana, Texas, Gulf of Mexico	Trunk
Texas Eastern Transmission Co[1,2]	Louisiana, Texas, Gulf of Mexico	Trunk
Texas Gas Transmission Co[1,2]	Louisiana, Gulf of Mexico	Trunk
Transcontinental Gas Pipeline Co[1,2]	Louisiana, Texas, Gulf of Mexico	Trunk
Trunkline Gas Co[1,2]	Louisiana, Texas, Gulf of Mexico	Trunk
Transwestern Pipeline Co	San Juan (CO,NM) & Permian Basins (TX)	Trunk
TransColorado Gas Transmission Co	New Mexico, Colorado (San Juan Basin)	Trunk
Trans-Union Interstate Pipeline Co	Arkansas, Louisiana	Trunk
West Texas Gas Co[2]	Southwest Texas	Trunk/Grid
Western Gas Interstate Co	Oklahoma, Texas	Trunk
Offshore to Onshore Pipelines		
Black Marlin Offshore Pipeline[3]	Offshore South Texas	Trunk
Blue Dolphin Pipeline Co[3]	Offshore South Texas	Trunk
Discovery Gas Transmission Co[3]	Offshore Eastern Louisiana	Trunk
Enbridge Offshore Pipelines	Offshore Western Louisiana	Trunk

(UTOS)[3]

High Island Offshore System[3]	Offshore Western Louisiana	Trunk
Matagorda Offshore Pipeline System[3]	Offshore Southeast Texas	Trunk
Nautilus Pipeline System[3]	Offshore Eastern Louisiana	Trunk
Sea Robin Pipeline Company[3]	Offshore Western Louisiana	Trunk
Stingray Pipeline System[3]	Offshore Western Louisiana	Trunk

*Offshore Gulf of Mexico Pipelines***

Anaconda Pipeline System	Green Canyon Area	Trunk
Canyon Chief Pipeline System	Mississippi Canyon Area	Trunk
Cleopatra Gathering System	Green Canyon Area (to Nautilus PL)	Trunk
Bluewater System	Offshore Central Louisiana	Trunk/Grid
East Breaks Gathering System	Alaminos Canyon Area	Trunk
Falcon Gas Pipeline System	Offshore Southeast Texas	Trunk
Garden Banks Gas Pipeline System[3]	Garden Banks Area	Trunk
Green Canyon System	Green Canyon Area	Trunk
Magnolia Gathering System	Green Canyon Area (to Garden Banks PL)	Trunk
Manta Ray Offshore Gathering System	Garden Banks Area (to Nautilus PL)	Trunk/Grid
Mississippi Canyon Gas Pipeline Co[3]	Mississippi Canyon Area	Trunk
Nemo Pipeline Co	Green Canyon Area	Trunk
Okeanos Deepwater System	Mississippi Canyon Area (to Discovery PL)	Trunk
Phoenix Gathering System	Vermillion Area	Trunk/Grid
Triton Gathering System	Garden Banks Area	Trunk
Typhoon Gas Gathering System	Green Canyon Area	Trunk
Viosca Knoll Gathering System	Offshore Eastern Louisiana	Trunk
Venice Gas Gathering System[3]	Offshore Eastern Louisiana	Trunk

*Intrastate Pipelines***

Acadian Gas Pipeline System	Southern Louisiana	Grid/Trunk
Arkansas Oklahoma Gas Co	Arkansas, Oklahoma	Trunk/Grid
Arkansas Western Pipeline Co	Arkansas	Trunk/Grid
Atmos Pipeline – Texas	East and West Texas	Trunk/Grid
Bridgeline Gas Systems	Southern Louisiana	Trunk/Grid
CCNG Transmission System	Southeast Texas	Trunk/Grid
Cypress Pipeline Co	Southern Louisiana	Trunk/Grid
Energy Transfer East Texas Pipeline	Northeastern Texas	Trunk
Enterprise Texas Intrastate Pipeline	Texas	Trunk/Grid
Enbridge Pipelines (North Texas)	North Texas	Trunk/Grid
Enbridge Pipelines (East Texas)	East Texas	Trunk/Grid
Enbridge Pipelines (Louisiana Intra)	Louisiana	Trunk/Grid
Enogex Pipeline System	Oklahoma	Trunk/Grid

Evangeline Gas Pipeline Co	Louisiana	Grid/Trunk
Guadalupe Pipeline Co	Texas Intrastate system	Trunk
Gulf Coast Pipeline System	Southeast Texas	Trunk/Grid
Houston Gas Pipeline Co	Southeast Texas	Trunk/Grid
Kinder Morgan Border Pipeline[2]	South Texas	Trunk
Kinder Morgan North Texas Pipeline	Northeastern Texas	Trunk
Kinder Morgan South Texas Pipeline	South Texas	Trunk
Kinder Morgan Texas Pipeline	Southeastern Texas	Trunk
Kinder Morgan Tejas Gas Pipeline	Southeastern Texas	Trunk
Louisiana Intrastate Gas (LIG) Co	Louisiana	Trunk/Grid
MarkWest Intrastate Pipeline Co	West Texas	Trunk
Norteno Pipeline Co[2]	South Texas	Trunk/Grid
Oasis Gas Pipeline Co	Central Texas	Trunk
Oklahoma Natural Gas Co	Oklahoma	Trunk/Grid
Public Service Co of New Mexico	New Mexico	Trunk/Grid
Regency Intrastate Gas Co	Southern Louisiana	Trunk/Grid
Southern Union Intrastate Pipelines Co	Oklahoma, Texas	Trunk/Grid
SouthWestern Energy Pipeline Co	Southern Arkansas	Trunk/Grid
Tidelands Pipeline System[2]	South Texas	Trunk/Grid
Vanderbuilt Pipeline System	Southeast Texas	Trunk/Grid
Westex Pipeline Co	West Texas	Trunk/Grid

*System Configuration - natural gas pipeline system design layout. Some natural gas systems are a combination of the trunk and grid. Where two are shown, the first represents the predominant system design.
 Trunk - systems are large-diameter long-distance trunklines that generally tie supply areas to natural gas market areas.
 Grid - systems are usually a network of many interconnections and delivery points that operate in and serve major natural gas market areas.
**Table is not necessarily inclusive of all intrastate natural gas pipelines operating in the region.
[1]Interstate natural gas pipeline that also has assets and operations in the Gulf of Mexico.
[2]Also provides natural gas transportation to and/or from Mexico.
[3]FERC jurisdictional system that is sometimes categorized as an offshore natural gas gathering system pipeline.
SOURCE: Energy Information Administration, Office of Oil & Gas.

Natural Gas Pipelines in the Central Region
Overview | Domestic Gas | Exports |
Regional Pipeline Companies & Links

Overview

Twenty interstate and at least twelve
intrastate natural gas pipeline companies
(see Table below) operate in the Central
Region (Colorado, Iowa, Kansas, Missouri,
Montana, Nebraska, North Dakota, South
Dakota, Utah, and Wyoming). Twelve
interstate natural gas pipeline systems enter
the region from the south and east while four
enter from the north carrying Canadian
supplies. The average utilization rates on
those shipping Canadian natural gas tend to
be higher than those carrying domestic
supplies.

Central Region Natural Gas Pipeline Network

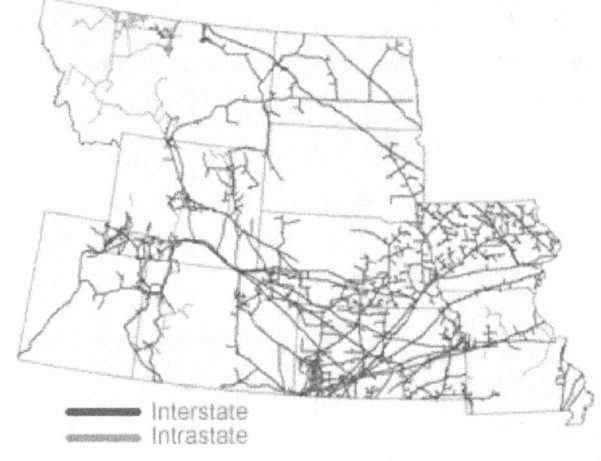

Interstate
Intrastate

The region produces more gas than it consumes (about 40 percent of its production) and therefore is a
net exporter of natural gas. The region has several large metropolitan markets that are major destinations
on the regional interstate natural gas pipeline network. Two of the largest are Denver, Colorado, served
by Colorado Interstate Gas Company, and Salt Lake City, Utah, which is served by Questar Pipeline
Company. Additional markets include the Kansas City metropolitan area of Kansas and Missouri, served
by the Southern Star Central Gas Pipeline Company (formerly Williams Gas Pipeline Central), KM
Interstate Gas Transmission Company, and Panhandle Eastern Pipeline Company systems; and the St
Louis, Missouri, area, which is served by the Centerpoint Mississippi River Transmission Company and
Southern Star Central Pipeline Company systems.

Transportation of Domestic Natural Gas Supplies

Several of the largest interstate natural gas pipelines that operate in the region retain much, if not all, of
their deliverability (capacity) for customers outside the region. For instance, the Kern River Gas
Transmission Company, which can receive up to 1.9 billion cubic feet (Bcf) per day in southwest
Wyoming, delivers more than 90 percent of its shipped volumes to destinations in Nevada and Southern
California. While some of this gas is consumed in the Central Region, most of the volumes are destined
for the Western Region of the country.

Even more extreme is the U.S. portion of the Alliance Pipeline Company system, which has only two
delivery points between the location at which it begins, the Saskatchewan/North Dakota border, and its
termination point in the vicinity of Joliet, Illinois.

Likewise, several large interstate natural gas pipelines originating in the Southwest (West Texas and
Oklahoma panhandles) provide only limited service within the region on their way to markets in the
Midwest. These natural gas pipeline systems flow northeastward primarily through Kansas, Missouri,
Nebraska and Iowa to reach Midwest natural gas markets. The two largest are the Natural Gas Pipeline
Company of America and the Panhandle Eastern Transmission systems. The ANR Pipeline Company
system follows a similar route, but it provides greater service within the region, especially in Iowa.

The Texas Eastern Transmission Company system only skirts the lower southeastern corner of the region, providing service at fewer than 15 delivery points. The Northern Natural Gas Pipeline Company system, on the other hand, has more than 225 delivery points within the Central Region, primarily in Iowa and Nebraska. While it has the capability to transport 1.6 Bcf per day into the region from the Southwest, it also receives large volumes of natural gas from interconnections to other interstate natural gas pipelines within the region. Indeed, when the Northern Natural Pipeline Company system exits the Central Region into the Midwest Region (Minnesota and Illinois), its local capacity has increased to more than 2.2 Bcf per day.

For the regional interstate natural gas pipelines, the largest State-to-State capacity within the region is Southern Star Central Gas Pipeline Company's line from Kansas to Missouri with a capacity of 1.3 Bcf per day. However, the average usage rates on this and similar service lines in the area are low, primarily owing to the seasonal nature of the service: low summer time flows tend to offset the high winter flows. In 2005, for instance, average annual utilization of Southern Star's Kansas/Missouri line was only about 21 percent.

In Colorado, the Rocky Mountain Natural Gas Company provides interconnections between natural gas producers in the western part of the State and the TransColorado Gas Transmission Company, which travels southward into New Mexico en route to the California market, through the Transwestern Gas Company and El Paso Natural Gas Pipeline Company systems.

In addition, in 2006, the TransColorado Gas Transmission Company installed a capability to reverse flow on its system, providing natural gas producers operating in the developing Unita/Piceance Basins of eastern Utah and western Colorado the opportunity to also transport their product eastward to the Cheyenne Hub in eastern Colorado where access to transportation services to Midwest markets is available. TransColorado Transmission also plans to further expand its northern flow capacity out of the San Juan Basin in southern Colorado in 2008 to give producers in that area greater access to these Midwest markets, in addition to their traditional Western regional market.

A major LDC in the western part of this region is Questar Gas Company (formerly Mountain Fuel Supply Company), which serves the Salt Lake City area and accounts for 99 percent of the receipts from interstate natural gas pipelines operating in Utah. Questar Pipeline Company, an affiliate, supplies the needs of this intrastate company.

The Public Service Company of Colorado is the major distributor of gas in Colorado, with more end-use customers in a single State than any other company in the region. Colorado Interstate Gas Company provides nearly all of the gas to this LDC.

Exports of Expanding Coal bed and Conventional Natural Gas Production

The expanding development of natural gas resources in Wyoming and Colorado, especially for coal bed methane and tight-sands natural gas, has greatly increased the amount of natural gas pipeline capacity built within and exiting the area in recent years. Since 1999, at least seven large-capacity header-laterals have been built in Wyoming alone, transporting natural gas from local gathering systems to interconnections with major interstate natural gas pipelines such as Colorado Interstate Gas Company, Wyoming Interstate Gas Company, Northwest Pipeline Company, Questar Overthrust Pipeline Company, and Questar Pipeline Company.

The new header-laterals include: the Fort Union Gathering header (0.6 Bcf per day), the Thunder Creek Gathering header (0.4 Bcf per day), the Bighorn Gas Gathering header (0.5 Bcf per day), Lost Creek Gathering header (0.3 Bcf per day), Jonah Gas Gathering system, (1.7 Bcf per day), and the Wyoming Interstate Pipeline Company's Medicine-Bow lateral (1.1 Bcf per day). In addition to these new routes, several existing intrastate natural gas pipelines have also provided support to producers needing natural gas transportation services in the area.

This new capacity, including recent expansions to the initial lines, represents an increase of 4.5 Bcf per day in take-away capability from the expanding Wyoming production basins. In addition, as this capacity has grown, other interstate natural gas pipelines, such as Colorado Interstate Gas Company, Trailblazer Pipeline Company, TransColorado Gas Transmission Company, and Kern River Gas Transmission Company, which provide transportation to more long-distance destinations, have also increased their capabilities as well.

Prior to 2004, the Trailblazer Pipeline Company, which receives natural gas from the Wyoming Interstate Company and Colorado Interstate Gas Company systems at the Cheyenne Hub in northeastern Colorado, provided the principal transportation route for Rocky Mountains area natural gas to reach markets in the Midwest Region (via interconnections with Natural Gas Pipeline Company of America and Northern Natural Gas Company in Nebraska). In 2004, however, the Colorado Interstate Gas Company completed its own new route (Cheyenne Plains Pipeline extension, 0.7 Bcf per day) from the Cheyenne hub to southern Kansas, giving it the capability to deliver natural gas to other interstate systems which have delivery points located within the Midwest Region.

Meanwhile, the Kern River Gas Transmission Company and TransColorado Gas Transmission Company expanded their capabilities to transport natural gas to western markets in recent years, with substantial increases in capacity from Utah and Nevada (Kern River Transmission Company, 1.8 Bcf per day) and the Piceance Basin of northwest Colorado and San Juan Basin of southern Colorado (TransColorado Gas Transmission Company, 0.7 Bcf per day).

In addition, several existing intrastate natural gas pipelines have provided support to producers needing natural gas transportation services out of the Rocky Mountains production areas. In Wyoming, these include the Overland Trails Pipeline Company system, which interconnects with the Questar Overthrust Pipeline Company and the Questar Pipeline Company systems in western Wyoming, and the MIGC and MGTC Pipeline Company systems which link the production fields in northeast Wyoming with the KM Interstate Gas Transmission Company system which, in turn, provides transportation southeastward into Nebraska.

New expansion capacity planned over the next several years in the Rocky Mountain area will support much of the growing natural gas production out in the area, primarily directing it to the Cheyenne Hub located in northeast Colorado. In addition to expansions on existing natural gas pipeline systems in the region, Kinder Morgan Energy Partners LP recently (2007) completed a new natural gas pipeline capable of moving up to 0.8 Bcf per day from the Piceance Basin of western Colorado to the Cheyenne Hub.

Complimenting this growth of capacity headed for the Cheyenne Hub, Kinder Morgan Energy Partners LP has also applied to the Federal Energy Commission for an approval to build a new 1.5 Bcf per day natural gas pipeline to take away additional supplies from the Cheyenne Hub to the Midwest and eventually to the Northeast Region. The first phase of this "Rockies Express Pipeline" project is scheduled for completion in early 2008.

Principal Natural Gas Pipeline Companies Serving the Central Region

Pipeline Name	Principal Supply Source(s)	System Configuration* Primary/Secondary
Interstate & Importing Pipelines		
Alliance Pipeline Co[1,2]	Canada	Trunk
ANR Pipeline Co[2]	Louisiana, Kansas, Texas	Trunk/Grid
Centerpoint Energy Pipeline Co	Kansas, Oklahoma, Texas	Trunk/Grid
Centerpoint Mississippi River Trans Co	Arkansas, Oklahoma	Trunk
Colorado Interstate Gas Co	Colorado, Oklahoma, Texas, Wyoming	Trunk/Grid
Kern River Gas Transmission Co[2]	Wyoming, Utah	Trunk

KM Interstate Gas Transmission Co	Kansas, Oklahoma, Wyoming	Trunk/Grid
MIGC Pipeline Co	Wyoming	Trunk
Natural Gas PL Co of America[2]	Kansas, Oklahoma, Louisiana, Texas	Trunk
Northern Border Pipeline Co[1,2]	Canada	Trunk
Northern Natural Gas Co	Kansas, Oklahoma, Texas	Trunk/Grid
Northwest Pipeline Co[2]	Green River (WY) & San Juan (CO) Basins	Trunk/Grid
Panhandle Eastern PL Co[2]	Kansas, Oklahoma, Texas	Trunk
Questar Pipeline Co	Wyoming, Colorado	Trunk/Grid
Questar Overthurst Pipeline Co	Wyoming	Trunk
TransColorado Gas Transmission Co[2]	Colorado	Trunk
Trailblazer Pipeline Co	Colorado, Wyoming	Trunk
Southern Star Central Pipeline Co	Kansas, Oklahoma, Wyoming	Trunk/Grid
Williston Basin Interstate PL Co[1]	Montana, Wyoming, Canada	Trunk/Grid
Wyoming Interstate Co	Wyoming	Trunk
*Intrastate Pipelines***		
Bighorn Gas Gathering Header (WY)	Wyoming	Trunk
Enbridge Pipelines (KPC) (KS,MO)	Interstate System	GridTrunk
Fort Union Gathering Header (WY)	Wyoming	Trunk
Harve Pipeline Co[1]	Canada	Trunk/Grid
Jonah Gathering System (WY)	Wyoming	Trunk
Lost Creek Gathering Co (WY)	Wyoming	Trunk
Missouri Gas Co (MO)	Interstate Pipelines	Grid
Missouri Public Service Co (MO)	Interstate Pipelines	Grid
NorthWestern Energy (MT)	Canada	Grid
Overland Trail Transmission Co (WY)	Wyoming	Trunk
Questar Gas Co (UT, WY)	Wyoming, Utah	Grid/Trunk
Rocky Mountain Natural Gas Co (CO)	Colorado	Trunk

*System Configuration - natural gas pipeline system design layout. Some systems are a combination of the trunk and grid. Where two are shown, the first represents the predominant system design.

Trunk - systems are large-diameter long-distance trunklines that generally tie supply areas to natural gas market areas.

Grid - systems are usually a network of many interconnections and delivery points that operate in and serve major natural gas market areas.

**Table is not necessarily inclusive of all intrastate natural gas pipelines operating in the region.

[1]Also operates natural gas import/export facilities located at the Canada border.

[2]Natural gas pipeline system transports all or a substantial portion of its deliveries beyond the Central Region.

SOURCE: Energy Information Administration, Office of Oil & Gas.

Natural Gas Pipelines in the Western Region

Overview | Transportation South | Transportation North |
Regional Pipeline Companies & Links

Western Region Natural Gas Pipeline Network

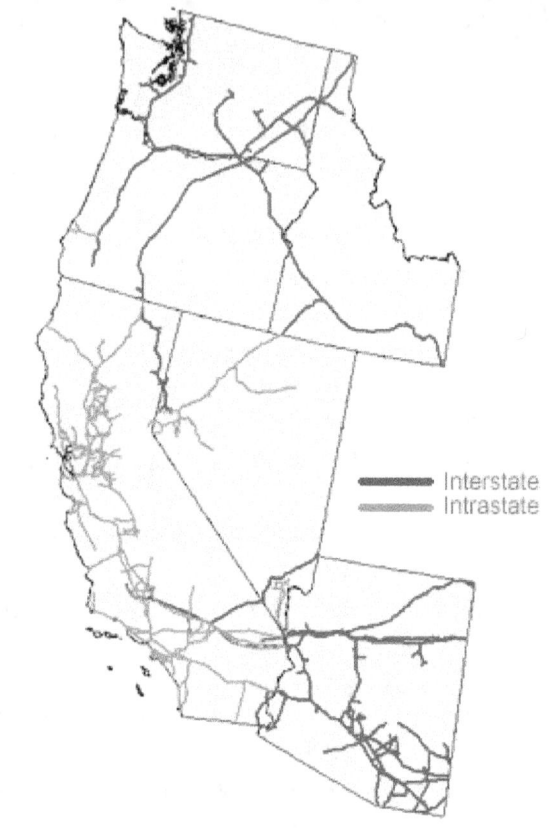

Overview

Ten interstate and nine intrastate natural gas pipeline companies provide transportation services to and within the Western Region (Arizona, California, Idaho, Nevada, Oregon, and Washington), the fewest number serving any region (see Table below). Slightly more than half the capacity entering the region is on natural gas pipeline systems that carry natural gas from the Rocky Mountain area and the Permian and San Juan basins.

These latter systems enter the region at the New Mexico-Arizona and Nevada-Utah State lines. The rest of the capacity arrives on natural gas pipelines that access Canadian natural gas at the Idaho and Washington State border crossings with British Columbia, Canada.

Natural Gas Transportation Services in the Southern Tier

The largest capacity natural gas pipeline within the region is the El Paso Natural Gas Company system. It has the capability to transport up to 6.2 billion cubic feet (Bcf) per day from natural gas production areas located in the Permian Basin of western Texas and the San Juan Basin of southern Colorado.

While the destination of a major portion of its deliveries is the California State border, this natural gas pipeline system also provides substantial service to customers in Arizona, especially to the growing natural gas fired electric power generation market. It is also the primary source of supply for the Southwest Gas Company (at the Arizona/Nevada State border), a major supplier of natural gas to southern Nevada and the Las Vegas metropolitan area.

Transwestern Pipeline Company's 1.3 Bcf per day natural gas pipeline system almost parallels the northern route of the El Paso Natural Gas Company system from West Texas through the San Juan Basin of northern New Mexico. It also delivers a large portion of its transported supplies to the California border and is a major participant within the Arizona marketplace.

Both the Transwestern Pipeline Company and El Paso Natural Gas Company systems deliver supplies to the three major intrastate natural gas pipelines operating in California: Southern California Gas Company (SoCal), California Gas Transmission Company (formerly PG&E Gas Transmission), and San Diego Gas & Electric Company (via the Southern California Gas Company system).

In addition, both Transwestern Pipeline Company and El Paso Natural Gas Company deliver to the Mojave Pipeline Company (0.4 Bcf per day) system, which enters the region at the northern Arizona/California border and crosses to Kern County, where it then merges with the Kern River Transmission Company system (1.8 Bcf per day). The Mojave Pipeline Company and Kern River Transmission Company systems (the latter begins at Opal, Wyoming, and extends through Utah and Nevada to Kern County, California) were the first interstate natural gas pipelines (in 1992) to extend into the State of California, which previously limited its territory to intrastate pipelines service only.

The El Paso Natural Gas Company's southern system is the principal deliverer of natural gas to the southern leg of the SoCal system at Blythe, California, which in turn provides a route for natural gas deliveries to the San Diego Gas & Electric Company system and natural gas exports to Mexico at several locations along the border. In late 2002, the El Paso Natural Gas Pipeline began deliveries to the new North Baja Pipeline Company system (0.5 Bcf per day), which has substantially increased the delivery of U.S. natural gas to electric power generation plants located in North Baja California, Mexico. [Map and Table of Import/Export Points] By late 2007, however, the North Baja Pipeline Company plans to install a bidirectional capability on its system which would permit future natural gas produced at LNG import facilities located in Baja, Mexico, to flow into the United States to customers in California and Arizona. The bi-directional capacity from Mexico to the United States will be expanded further by 2010, increasing the capacity from Mexico to as high as 2.0 Bcf per day.

In July 2002, Questar's Southern Trails Pipeline Company system, a converted oil pipeline with a capacity of 0.1 Bcf per day, began transporting natural gas from the four-corners area of Utah/Arizona/New Mexico/Colorado to the California/Arizona border near Needles. Originally designed to extend into California as far as the Pacific Coast in the Long Beach area, the final leg of the project was canceled owing to California's anti-bypass regulatory policy. Although the system does include interconnections with California Gas Transmission Company and the Southern California Gas Company within the State of California near the Arizona/California border, for the time being, the final destination for much of the gas transported on the Southern Trails Pipeline is a natural gas fired electric power generation plant located in western Arizona.

Transportation Services in the Northern Tier

While the California Gas Transmission Company system receives about 40 percent of its requirements from El Paso Natural Gas and Transwestern Gas at Topock, Arizona, it receives the bulk of its supplies (about 60 percent) at its northern terminus (2.2 Bcf per day) with the Gas Transmission Northwest interstate natural gas pipeline, at the Oregon/California border. The Gas Transmission Northwest Company system transports Canadian natural gas from the Canada/Idaho border through Washington State and Oregon. At the same location (Malin, Oregon) where the Gas Transmission Northwest Company system delivers into the California Gas Transmission Company system, it also delivers into the Tuscarora Pipeline Company (0.1 Bcf per day) system, which extends natural gas transportation from the northern California border to the Reno, Nevada, area.

The Gas Transmission Northwest Company system also interconnects with Northwest Pipeline Company at Stansfield, Oregon, providing Canadian shippers that wish to deliver natural gas (up to 0.5 Bcf per day) to the Vancouver, British Columbia, area via an alternative route to the existing northern natural gas pipeline route within Canada, an opportunity to do so. Initiated in 1999, this routing of Canadian natural gas was the first time natural gas export service was provided by the Northwest Pipeline Company at Sumas, Washington, which was normally the import point for Canadian supplies destined for Washington State and northern Oregon (via Northwest Natural Gas Company). Sumas is the northern terminus for the Northwest Pipeline Company system, where it receives up to 1.4 Bcf per day of Canadian supplies and transports them within the northern tier of the region. Several local natural gas importing pipelines (Ferndale Pipeline Company and Sumas Energy International Company) also operate though the Sumas location. [Map and Table of Import/Export Points]

The Northwest Pipeline Company system begins at Sumas, Washington, and extends southeast through Oregon, Idaho, northern Utah, Wyoming, and southward into the San Juan Basin in southern Colorado. In addition to delivering Canadian natural gas along the northern section in Washington, Oregon and Idaho, the Northwest system is bidirectional, with the capability to direct natural gas supplies from the prolific Wyoming natural gas fields and the San Juan Basin to these northwest States when needed. Northwest Pipeline Company also is the only source of supply for the Pauite Pipeline Company system (0.2 Bcf per day) which taps the Northwest system at the Idaho/Nevada border and delivers natural gas to the Reno, Nevada area from the northeast (while the Tuscarora Pipeline Company delivers from northern California).

Principal Natural Gas Pipeline Companies Serving the Western Region

Pipeline Name	Principal Supply Source(s)	System Configuration* Primary/Secondary
Interstate & Importing Pipelines		
El Paso Natural Gas Co[1]	San Juan (CO,NM) & Permian Basin (TX)	Trunk/Grid
Gas Transmission Northwest[1]	Canada	Trunk
Kelso-Beaver Pipeline Co	Interstate System	Trunk
Kern River Transmission Co	Wyoming, Utah	Trunk
Mojave Pipeline Co	Southwest (via the Interstate System)	Trunk
North Baja Pipeline Co	Southwest (via the Interstate System)	Trunk
Northwest Pipeline Co[1]	Canada, Green River (WY) & San Juan (CO) Basins	Trunk/Grid
Southern Trails Pipeline (Questar)	San Juan Basin (CO,NM)	Trunk
Transwestern Pipeline Co	San Juan (CO,NM) & Permian Basins (TX)	Trunk
Tuscarora Gas Transmission Co	Canada (via Interstate System)	Trunk
*Intrastate Pipelines**		
Cascade Natural Gas Co	Interstate System	Grid/Trunk
Ferndale Pipeline Co[2]	Canada	Trunk
Northwest Natural Gas Co	Interstate System	Trunk/Grid
Pauite Pipeline Co (NV)	Interstate System	Trunk
California Gas Transmission Co	Interstate System & California Production	Trunk/Grid
San Diego Gas & Electric Co (CA)[1]	Intrastate System	Trunk/Grid
Southern California Gas Co (CA)[1]	Interstate System & California Production	Trunk/Grid
Southwest Gas Co (CA, NV)	Interstate System	Trunk/Grid
Sumas International Pipeline[3]	Canada/Northwest Pipeline Co	Trunk

*System Configuration - natural gas pipeline system design layout. Some systems are a combination of the trunk and grid. Where two are shown, the first represents the predominant system design.
 Trunk - systems are large-diameter long-distance trunklines that generally tie supply areas to natural gas market areas.
 Grid - systems are usually a network of many interconnections and delivery points that operate in and serve major natural gas market areas.
**Table is not necessarily inclusive of all intrastate natural gas pipelines operating in the region
[1]Also operates natural gas import/export facilities located at the Canada or Mexico border.
[2]Short haul natural gas pipeline that imports Canadian natural gas for local industrial use.
[3]Facilities were sold to Northwest Pipeline Co in 2004.
SOURCE: Energy Information Administration, Office of Oil & Gas.

Natural Gas Pipelines - Transporting Natural Gas

Intrastate Natural Gas Pipeline Segment

Overview

Intrastate natural gas pipelines operate within State borders and link natural gas producers to local markets and to the interstate pipeline network. Approximately 29 percent of the total miles of natural gas pipeline in the U.S. are intrastate pipelines.

Although an intrastate pipeline system is defined as one that operates totally within a State, an intrastate pipeline company may have operations in more than one State. As long as these operations are separate, that is, they do not physically interconnect, they are considered intrastate, and are not jurisdictional to the Federal Energy Regulatory Commission (FERC). More than 90 intrastate natural gas pipelines operate in the lower-48 States.

Texas has more intrastate natural gas pipeline miles (over 43,000) than any other State.

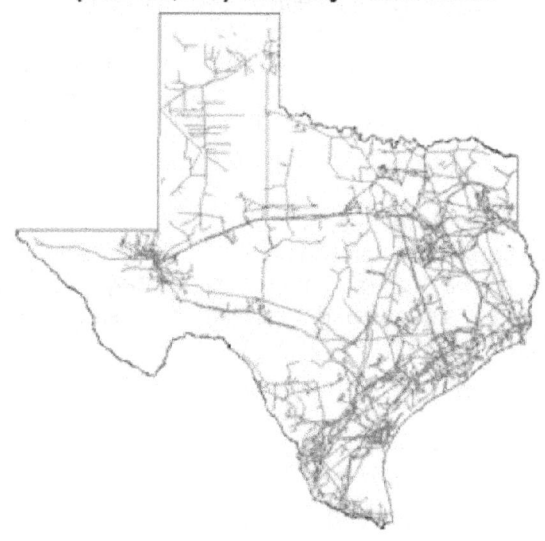

More information related to intrastate pipelines…

Intrastate Natural Gas Pipeline Companies – spreadsheet

Selected State Highlights ---

- **Texas:**
    - Texas is the top ranked natural gas consuming State.
    - Intrastate pipelines in Texas account for 43,000 of the 57,000 miles of natural gas pipelines in the State.
    - The largest intrastate pipelines in Texas are Enterprise Texas Pipeline Company (7,500 miles) and the Atmos Pipeline-Texas Company (6,200 miles).
    - The intrastate network in Texas has experienced significant growth over the past several years as a result of increased demand for pipeline capacity caused by the rapid development and expansion of natural gas production in the Barnett Shale Formation.
    - New pipelines have been built, and expansions to existing ones undertaken, to meet increased demand.
- **California:**
    - California is ranked the second largest natural gas consuming State.
    - Intrastate transportation and distribution are dominated by California Gas Transmission Co. (PG&E) (3,477 miles), Southern California Gas (SoCal) Company (1,887 miles), and the San Diego Gas and Electric Company.
    - SoCal and PG&E are two of the largest distribution companies in the entire United States.

- **Wyoming, Colorado, and Utah:**
  - Development of new, large-diameter intrastate gathering pipelines is proceeding at a fast pace in this area, as proved reserves of coalbed methane, tight sands, and conventional natural gas supplies are being developed.
  - During the past several years, at least 7 large-capacity pipeline header systems have been built in Wyoming to transport natural gas from local gathering systems.
  - In western Colorado and eastern Utah, several new gathering systems have been developed to feed gas into the interstate pipeline network.
  - The Overland Trails Transmission Company and the Rocky Mountain Natural Gas Company are the principal intrastate pipelines in this area, and provide some of the primary links between expanding natural gas production fields in the area and the interstate pipeline network.

In some instances, an intrastate natural gas pipeline may also be classified as a "Hinshaw" pipeline. Although such pipelines receive all of their supplies from interstate pipeline sources, and therefore fall within FERC's regulatory purview, they have been exempted from its jurisdiction because the gas they deliver is consumed totally within the state in which they operate.

About U.S. Natural Gas Pipelines - Transporting Natural Gas

Network Configuration and System Design

Overview | Transmission/Storage | Design Criteria | Importance of Storage| Overall Pipeline System Configuration

Overview

A principal requirement of the natural gas transmission system is that it be capable of meeting the peak demand of its shippers who have contracts for firm service.

To meet this requirement, the facilities developed by the natural gas transmission industry are a combination of transmission pipelines to bring the gas to the market areas and of underground natural gas storage sites and liquefied natural gas (LNG) peaking facilities located in the market areas.

Generalized Natural Gas Pipeline Capacity Design Schematic

click to enlarge

More information related to network configuration ...

States Dependent on Interstate Pipelines - map
Region to Region System Capacity Levels - map
Major Transportation Corridors - map

Major Interstate Pipeline Companies – Appendix table
Pipeline Mileage by State & Region – Appendix table

Sizes of Transmission Lines and Integrated Storage Sites

The design of natural gas transmission pipelines and integrated storage sites represents a balance of the most efficient and economical mix of delivery techniques given the operational requirements facing the pipeline company, the number and types of transportation customers, and available access to supplies from production areas or from underground storage.

Many natural gas pipeline systems are configured principally for the long-distance transmission of natural gas from production regions to market areas. These long-distance systems are often referred to as trunklines.

At the other extreme are the grid systems, which generally operate in and serve major market areas. Many of the grid systems can be categorized as regional distribution systems. For the most part, they receive their supplies of natural gas from the major trunklines or directly from local production areas. The grid systems transport natural gas to local distribution companies and large-volume consumers.

Design Criteria and Pipeline Size

The design process includes the development of cost estimates for various possible combinations of pipe size, compression equipment, and inter-station distances to find the optimal combination that minimizes the transportation cost, given the desired flexibility and expandability goals.

New trunklines typically are built with a larger diameter pipe than will be needed initially but with compression capacity limited to meeting current needs. Compressors can be added, in either new or existing stations, to increase capacity as growth in load occurs.

A number of factors are involved in calculating how much natural gas a pipeline can carry. However, the most important factors are the diameter of the pipe and its operating pressure.

Standard design codes require that all pipelines passing through populated areas reduce its maximum operating pressures for safety reasons.

It had become common practice to maintain nominal pipe diameter but increase wall thickness where a line had to be derated for its surroundings (change in external stresses due to earth or traffic loads) in order to keep the working pressure rating more constant along the line. Increasing the pipe wall thickness or strength of the pipe will enable the pipe to withstand a greater pressure between operating and design pressure to adhere to safety requirements.

Importance of Underground Storage Integration

Underground storage is an essential component of an efficient and reliable interstate natural gas transmission and distribution network. The size and profile of the transmission system often depends in part on the availability of storage.

Access to underground natural gas storage facilities, particularly those located in consuming areas, permits the mainline transmission pipeline operator to design the portion of its system located upstream of storage facilities to accommodate the level of total shipper firm (reserved) capacity commitments and the pipeline operator's potential storage injection needs, commonly referred to as "baseload" requirements.

The portion of the transmission system located downstream of the storage area (including LNG peaking facilities) is designed to accommodate the maximum peak-period requirements of shippers, local distribution companies, and consumers in the area. It is generally sized to reflect the total peak-day withdrawal (deliverability) level of all storage facilities linked to the system and estimated potential peak-period demand requirements.

The daily deliverability from storage can also be factored into the design needs of a new pipeline or the expansion needs of an existing one. Some underground storage facilities are located in production areas at the beginning of the pipeline corridor and, in contrast to storage near consuming markets, can be used to store gas that may not be marketable at the time of production.

For instance, natural gas produced in association with oil production is a function of oil market decisions, which may not coincide with natural gas demand or available pipeline capacity to transport the gas to end-use markets. Another example is the storage of natural gas produced from low-pressure wells, which may be injected into storage during the off-peak season and delivered, at high pressure, to the mainline during the peak season.

These sites can be used by shippers to store short-term incremental supplies that exceed their reserve capacity on the pipeline system and the reverse when supplies fall below reserved capacity. Thus, the pipeline is relieved of additional demands for capacity brought on by temporary swings in the transportation demands of its customers.

Overall Pipeline System Configuration

The overall pipeline system configuration should result in a comparatively lower usage level (load factor) for downstream facilities in the summer season but a much higher, albeit shorter term, usage level during the peak-demand season. The upstream trunkline portion of the system, on the other hand, could operate at a more sustained high load factor throughout the year. (This design minimizing is oftentimes referred to as peak-shaving.)

With underground natural gas storage and LNG peaking facilities configured into a natural gas pipeline system, especially one serving climate-sensitive markets such as the Midwest and Northeast, system operators can minimize the facilities and costs involved in building the "trunkline" portion of their system. Natural gas shippers, on the other hand, could avoid unnecessary costs incurred if they reserved additional firm capacity on an entire transmission system, rather than only a portion that would be used only on a few days during the winter season.

During the nonheating season, for instance, when shippers do not need all the contracted capacity to meet their customer's current consumption requirements, natural gas can be transported and injected into storage. By the beginning of the heating season (November 1), inventory levels are generally at their annual peak. Working gas, the portion of natural gas in storage sites available for withdrawal and delivery to markets, is then withdrawn during periods of peak demand.

In addition, the pipeline company can avoid the need to expand transmission capacity from production areas by using existing, or establishing new storage facilities in market areas where there is a strong seasonal variation in demand and where the system may be subjected to operational imbalances.

Natural Gas Pipeline Capacity & Utilization

Overview | Utilization Rates | Integration of Storage | Varying Rates of Utilization | Measures of Utilization

Overview of Pipeline Utilization

Natural gas pipeline companies prefer to operate their systems as close to full capacity as possible to maximize their revenues. However, the average utilization rate (flow relative to design capacity) of a natural gas pipeline system seldom reaches 100%. Factors that contribute to outages include:

- Scheduled or unscheduled maintenance
- Temporary decreases in market demand
- Weather-related limitations to operations

Most companies try to schedule maintenance in the summer months when demands on pipeline capacity tend to be lower, but an occasional unanticipated incident may occur that suspends transmission service.

Interregional Transmission Pipeline Capacity Levels

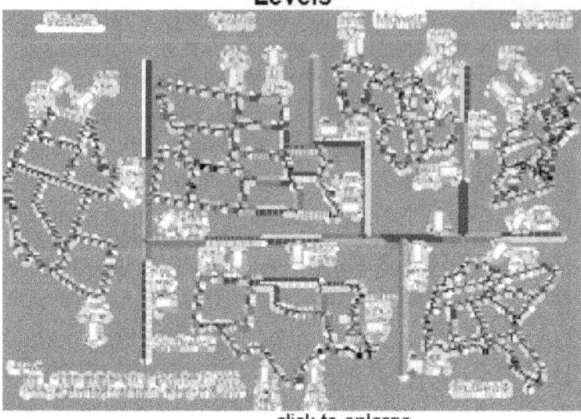

click to enlarge

More information related to pipeline capacity and utilization…

States Dependent on Interstate Pipelines - map
Major Transportation Corridors - map

Major Interstate Pipeline Companies – Appendix table
Pipeline Mileage by State & Region – Appendix table

Interstate Pipeline Capacity on a State-to-State Level - spreadsheet

Utilization Rates

Utilization rates below 100% do not necessarily imply that additional capacity is available for use. A pipeline company that primarily serves a seasonal market, for instance, may have a relatively low average utilization rate especially during the summer months. But that does not mean there is unreserved capacity on a long-term basis.

On the other hand, during periods of high demand for natural gas transportation services, usage on some portions of a pipeline system may exceed 100% of certificated capacity. Certificated capacity represents a minimum level of service that can be maintained over an extended period of time, and not the maximum throughput capability of a system or segment on any given day.

Exceeding 100% of capacity is accomplished by secondary compression and/or line packing, which means that compression is increased, within safety limits, to raise throughput temporarily.

Integration of Storage Capacity

Integrating storage capacity into the natural gas pipeline network design can increase average-day utilization rates. This integration involves moving not only natural gas currently being produced but natural gas that has been produced earlier and kept in temporary storage facilities.

Storage is usually integrated into or available to the system at the production and/or consuming end as a means of balancing flow levels throughout the year. Trunklines serving markets with significant storage capacity have greater potential for achieving a high utilization rate because the load moving on these pipelines can be leveled. To the extent that these pipelines serve multiple markets, they also can achieve higher utilization rates because of the load diversity of the markets they serve.

Varying Rates of Utilization

Trunklines, which are generally upstream (closer to) the natural gas production fields and storage areas, may sometimes exhibit peak period utilization rates exceeding 100% because they are occasionally capable of handling much larger volumes than indicated by the operational design certificated by FERC.

Utilization on the grid systems, which are closer to the consuming market areas and downstream of the storage fields, is more likely to reflect a seasonal load profile of the market being served. The grid-type systems usually operate at lower average utilization levels than trunklines and usually show marked variation between high and low flow levels, reflecting seasonal service and local market characteristics.

Measures of Pipeline Utilization

There are several ways that natural gas pipeline system utilization may be estimated, as demonstrated in the following cases:

- As a measure of the average-day natural gas throughput relative to estimates of system capacity at State and regional boundaries
- The systemwide pipeline flow rate, which highlights variations in system usage relative to an estimated system peak throughput level
- A system peak-day usage rate, which generally reflects peak system deliveries relative to estimated system capacity

The latter measure is a good indication of how well the design of the system matches current shipper peak-day needs. For example, when a pipeline shows a comparatively low average usage rate (based on annual or monthly data) yet shows a usage rate approaching 100 percent on its peak day, it indicates that the system is called upon and is capable of meeting its shipper's maximum daily needs. Nevertheless, a large spread between average usage rates and peak-day usage rates may indicate opportunities to find better ways to utilize off-peak unused capacity.

In some cases, utilization rates exceeding 100 percent may be an artifact of the data that obscures the true operational status of the pipeline. In some instances the sum of individual transportation transactions may exceed pipeline capacity even though physically the pipeline may not be full. For example, suppose a segment from points A to D (with points B and C between A and D) has a capacity of 200 million cubic feet (MMcf) per day. Suppose further that this segment handles a 100 MMcf per day transaction from A to B, a second of 100 MMcf per day from B to C, and a third of 100 MMcf per day from C to D. The pipeline company will report transportation volumes of 300 MMcf per day, even though its capacity is 200 MMcf per day but is only 50 percent utilized on any one segment.

U.S. Natural Gas Regulatory Authorities

Beginning | Regulations Today | Coordinating Agencies | Regulation of Mergers and Acquisitions

Beginning of Industry Restructuring

In April 1992, the Federal Energy Regulatory Commission (FERC) issued its Order 636 and transformed the interstate natural gas transportation segment of the industry forever. Under it, interstate natural gas pipeline companies were required to restructure their operations by November 1993 and split-off any non-regulated merchant (sales) functions from their regulated transportation functions.

This new requirement meant that interstate natural gas pipeline companies were allowed to only transport natural gas for their customers. The restructuring process and subsequent operations have been supervised closely by FERC and have led to extensive changes throughout the interstate natural gas transportation segment which have impacted other segments of the industry as well.

Regulations Today

Most natural gas pipelines in the United States, including many in the intrastate segment as well, now only transport natural gas and no longer buy and sell it. Although interstate natural gas pipelines are no longer subject to as much regulation as before Order 636, many aspects of their operations and business practices, are still subject to regulatory oversight.

For example, FERC determines the rate-setting methods for interstate pipeline companies, sets rules for business practices, and has the sole responsibility for authorizing the siting, construction, and operations of interstate pipelines, natural gas storage fields, and liquefied natural gas (LNG) facilities.

Agencies with an interest in natural gas transportation developments ...

Federal Energy Regulatory Commission (FERC)
Department of Energy's Office of Fossil Energy
Environmental Protection Agency
Department of Transportation's Office of Pipeline Safety (OPS).
Department of Justice (DOJ)
Federal Trade Commission (FTC)
Internal Revenue Service (IRS)
Nuclear Regulatory Commission (NRC).
Department of Homeland Security's U.S. Coast Guard
Department of Interior's Bureau of Land Management
Department of Interior's Maritime Administration
Department of Interior's Bureau of Indian Affairs
Department of Interior's U.S. Geological Survey
Department of Agriculture's U.S. Forest Service

State Utility Commissions

Regulatory bodies have the authority to suspend some rules and regulations under specific circumstances, especially in response to emergency and disaster situations, placing needed projects on a regulatory fast-track.

Coordinating with other Regulatory Agencies

Almost all applications to FERC for interstate natural gas pipeline projects require some level of coordination with one or more other Federal agencies. For example, the Environmental Protection Agency assists FERC and/or State authorities in determining if the environmental aspects of a pipeline development project meet acceptable guidelines. FERC is also required to take the lead on the environmental reviews under the National Environmental Policy Act, the Endangered Species Act, the National Historic Preservation Act, and the Magnuson-Stevens Act.

Governing the safety standards, procedures, and actual development and expansion of any pipeline system is the job of the U.S. Department of Transportation's Office of Pipeline Safety (OPS). A pipeline may not begin operations until a line, or line segment, has been certified safe by the OPS. The OPS retains jurisdiction for safety over the lifetime of the pipeline.

Regulation of Mergers and Acquisitions

To help ensure fairness and to preserve open markets, agencies at the Federal, State, and sometimes local levels examine mergers and acquisitions. Among those most actively involved in examining mergers and acquisitions at the Federal level are FERC, the Department of Justice, the Federal Trade Commission, the Internal Revenue Service, and the Nuclear Regulatory Commission. State public utility commissions, or their equivalent, also have responsibility for oversight in mergers and acquisitions of pipeline companies.

Each of the various agencies has the power to impose conditions that must be met to get approval for a merger or acquisition. If these conditions are not satisfied, the agencies can prevent the corporate combination from taking place. For example, analysis of mergers or acquisitions for potential harm to the consumer is under the shared jurisdiction of the Federal Trade Commission and the Department of Justice, where the concept of market power plays a central role in the antitrust review process.

Transportation Process and Flow

Overview | Gathering System | Processing Plant | Transmission Grid | Market Centers/Hubs | Underground Storage | Peak Shaving

Overview

Transporting natural gas from the wellhead to the final customer involves several physical transfers of custody and multiple processing steps. A natural gas pipeline system begins at the natural gas producing well or field. Once the gas leaves the producing well, a pipeline gathering system directs the flow either to a natural gas processing plant or directly to the mainline transmission grid, depending upon the initial quality of the wellhead product.

The processing plant produces pipeline-quality natural gas. This gas is then transported by pipeline to consumers or is put into underground storage for future use. Storage helps to maintain pipeline system operational integrity and/or to meet customer requirements during peak-usage periods.

Natural Gas Transmission Path

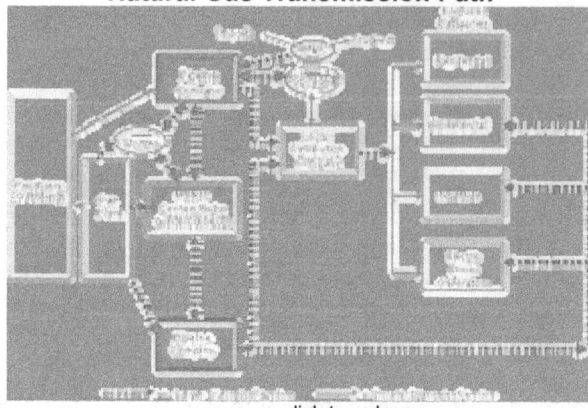

click to enlarge

More information related to process and flow:

U.S. Natural Gas Pipeline Compressor Stations - map
Major North American Natural Gas Market Centers & Hubs - map

Natural Gas Processing Plants: 1995 & 2004 - report
Natural Gas Market Centers and Hubs, 2003 - report
Underground NG Storage Expansions: 1998-2005 - report

Transporting natural gas from wellhead to market involves a series of processes and an array of physical facilities. Among these are:

- **Gathering Lines** – These small-diameter pipelines move natural gas from the wellhead to the natural gas processing plant or to an interconnection with a larger mainline pipeline.
- **Processing Plant** – This operation extracts natural gas liquids and impurities from the natural gas stream.
- **Mainline Transmission Systems** – These wide-diameter, long-distance pipelines transport natural gas from the producing area to market areas.
- **Market Hubs/Centers** – Locations where pipelines intersect and flows are transferred.
- **Underground Storage Facilities** – Natural gas is stored in depleted oil and gas reservoirs, aquifers, and salt caverns for future use.
- **Peak Shaving** – System design methodology permitting a natural gas pipeline to meet short-term surges in customer demands with minimal infrastructure. Peaks can be handled by using gas from storage or by short-term line-packing.

The Natural Gas Gathering System

A natural gas pipeline system begins at a natural gas producing well or field. In the producing area many of the pipeline systems are primarily involved in "gathering" operations. That is, a pipeline is connected to a producing well, converging with pipes from other wells where the natural gas stream may be subjected to an extraction process to remove water and other impurities if needed. Natural gas exiting the production field is usually referred to as "wet" natural gas if it still contains significant amounts of hydrocarbon liquids and contaminants.

Under certain conditions some or all of the natural gas produced at a well may be returned to the reservoir in cycling, repressuring, or conservation operations and/or vented and flared. At this stage it is a mixture of methane and other hydrocarbons, as well as some non-hydrocarbons, existing in the gaseous phase or in a solution with crude oil. The principal hydrocarbons normally contained in the natural gas mixture are methane, ethane, propane, butane, and pentane. Typical non-hydrocarbon gases that may be present in reservoir natural gas are water vapor, carbon dioxide, helium, hydrogen sulfide, and nitrogen.

In proximity to the well are facilities that produce what is referred to as "lease condensate", that is, a mixture consisting primarily of pentanes and heavier hydrocarbons which is recovered as a liquid from natural gas. Other natural gas liquids, such as butane and propane, are recovered at downstream natural gas processing plants or facilities (see below).

Once it leaves the producing area, a pipeline system directs flow either to a natural gas processing plant or directly to the mainline transmission grid. Nonassociated natural gas, that is, natural gas that is not in contact with significant quantities of crude oil in the reservoir, is sometimes of pipeline quality after undergoing a decontamination process in the production area, and does not need to flow through a processing plant prior to entering the mainline transmission system.

The Natural Gas Processing Plant

The principal service provided by a natural gas processing plant to the natural gas mainline transmission network is that it produces pipeline quality natural gas. Natural gas mainline transmission systems are designed to operate within certain tolerances. Natural gas entering the system that is not within certain specific gravities, pressures, Btu content range, or water content level will cause operational problems, pipeline deterioration, or even cause pipeline rupture.

Natural gas processing plants are also facilities designed to recover natural gas liquids from a stream of natural gas that may or may not have passed through lease separators and/or field separation facilities. These facilities also control the quality of the natural gas to be marketed. Several types of natural gas processing plants, employing various techniques and technologies to extract contaminants and natural gas liquids, are used to produce pipeline quality "dry" gas. At many processing plants the primary objective is the production of dry gas (demethanizing). Any remaining natural gas liquids extraction stream is directed to a separate plant to undergo what is referred to as a "fractionation" process.

But a number of natural gas processing plants do include these fractionation facilities, where saturated hydrocarbons are removed from natural gas and separated into distinct parts, or "fractions," such as propane, butane, and ethane. Essentially, natural gas is methane, a colorless, odorless, flammable hydrocarbon gas (CH_4). Also present in natural gas production, especially that in association with oil production, are a number of petroleum gases. They include (in addition to ethane, propane and butane) ethylene, propylene, butylene, isobutane, and isobutylene. They are derived from crude oil refining or natural gas fractionation and are liquefied through pressurization.

The Transmission Grid and Compressor Stations

The natural gas mainline (transmission line) is a wide-diameter, often-times long-distance, portion of a natural gas pipeline system, excluding laterals, located between the gathering system (production area), natural gas processing plant, other receipt points, and the principal customer service area(s). The lateral, usually of smaller diameter, branches off the mainline natural gas pipeline to connect with or serve a specific customer or group of customers.

A natural gas mainline system will tend to be designed as either a grid or a trunkline system. The latter is usually a long-distance, wide-diameter pipeline system that generally links a major supply source with a market area or with a large pipeline/LDC serving a market area. Trunklines tend to have fewer receipt points (usually at the beginning of its route), fewer delivery points, interconnections with other pipelines, and associated lateral lines.

A grid type transmission system is usually characterized by a large number of laterals or branches from the mainline, which tend to form a network of integrated receipt, delivery and pipeline interconnections that operate in, and serve major market areas. In form, they are similar to a local distribution company (LDC) network configuration, but on a much larger scale.

Between the producing area, or supply source, and the market area, a number of compressor stations are located along the transmission system. These stations contain one or more compressor units whose purpose is to receive the transmission flow (which has decreased in pressure since the previous compressor station) at an intake point, increase the pressure and rate of flow, and thus, maintain the movement of natural gas along the pipeline.

Compressor units that are used on a natural gas mainline transmission system are usually rated at 1,000 horsepower or more and are of the centrifugal (turbine) or reciprocating (piston) type. The larger compressor stations may have as many as 10-16 units with an overall horsepower rating of from 50,000 to 80,000 HP and a throughput capacity exceeding three billion cubic feet of natural gas per day. Most compressor units operate on natural gas (extracted from the pipeline flow); but in recent years, and mainly for environmental reasons, the use of electricity driven compressor units has been growing.

Many of the larger mainline transmission routes are what is generally referred to as "looped." Looping is when one pipeline is laid parallel to another and is often used as a way to increase capacity along a right-of-way beyond what is possible on one line, or an expansion of an existing pipeline(s). These lines are connected to move a larger flow along a single segment of the pipeline system. Some very large pipeline systems have 5 or 6 large diameter pipes laid along the same right-of-way. Looped pipes may extend the distance between compressor stations, where they can transfer part of their flow, or the looping may be limited to only a portion of the line between stations. In the latter case, the looping often serves as essentially a storage device, where natural gas can be line-packed as a way to increase deliveries to local customers during certain peak periods.

To address the potential for pipeline rupture, safety cutoff meters are installed along a mainline transmission system route. Devices located at strategic points are designed to detect a drop in pressure that would result from a downstream or upstream pipeline rupture and automatically stop the flow of natural gas beyond its location. Monitoring the pipeline as a whole are apparatus known as (SCADA Systems Control and Data Acquisition) systems. SCADA systems provide monitoring staff the ability to direct and control pipeline flows, maintaining pipeline integrity and pressures as natural gas is received and delivered along numerous points on the system, including flows into and out of storage facilities.

Natural Gas Market Centers/Hubs

Natural gas market centers and hubs evolved, beginning in the late 1980s, as an outgrowth of natural gas market restructuring and the execution of a number of Federal Energy Regulatory Commission's (FERC) Orders culminating in Order 636 issued in 1992. Order 636 mandated that interstate natural gas pipeline companies transform themselves from buyers and sellers of natural gas to strictly natural gas transporters. Market centers and hubs were developed to provide new natural gas shippers with many of the physical capabilities and administrative support services formally handled by the interstate pipeline company as "bundled" sales services.

Two key services offered by market centers/hubs are transportation between and interconnections with other pipelines and the physical coverage of short-term receipt/delivery balancing needs. Many of these centers also provide unique services that help expedite and improve the natural gas transportation process overall, such as Internet-based access to natural gas trading platforms and capacity release programs. Most also provide title transfer services between parties that buy, sell, or move their natural gas through the center.

As of the end of 2004, there were a total of 37 operational market centers in the United States (28) and Canada (9).

Underground Storage Facilities

At the end of the mainline transmission system, and sometimes at its beginning and in between, underground natural gas storage and LNG (liquefied natural gas) facilities provide for inventory management, supply backup, and the access to natural gas to maintain the balance of the system. There are three principal types of underground storage sites used in the United States today: depleted reservoirs in oil and/or gas fields, aquifers, and salt cavern formations. In one or two cases mine caverns have been used. Two of the most important characteristics of an underground storage reservoir are the capability to hold natural gas for future use, and the rate at which natural gas inventory can be injected and withdrawn (its deliverability rate).

Most underground storage facilities, 323 out of 398 at the beginning of 2007, are depleted reservoirs, which are close to consumption centers and which were relatively easy to convert to storage service. In some areas, however, most notably the Midwestern United States, some natural aquifers have been converted to natural gas storage reservoirs. An aquifer is suitable for natural gas storage if the water-bearing sedimentary rock formation is overlaid with an impermeable cap rock. While the geology of aquifers is similar to that of depleted production fields, their use in natural gas storage usually requires more base (cushion) gas and greater monitoring of withdrawal and injection performance. Deliverability rates may be enhanced by the presence of an active water drive.

During the past 20 years, the number of salt cavern storage sites has grown significantly because of its rapid cycling (inventory turnover) capability coupled with its ability to respond to daily, even hourly, variations in customer needs. The large majority of salt cavern storage facilities have been developed in salt dome formations located in the Gulf Coast States. Salt caverns leached from bedded salt formations in Northeastern, Midwestern, and Western States have also been developed but the number has been limited due to a lack of suitable geology. Cavern construction is more costly than depleted field conversions when measured on the basis of dollars per thousand cubic feet of working gas capacity, but the ability to perform several withdrawal and injection cycles each year reduces the per-unit cost of each thousand cubic feet of natural gas injected and withdrawn.

Peak Shaving

Underground natural gas storage inventories provide suppliers with the means to meet peak customer requirements up to a point. Beyond that point the distribution system still must be capable of meeting customer short-term peaking and volatile swing demands that occur on a daily and even hourly basis. During periods of extreme usage, peaking facilities, as well as other sources of temporary storage, are relied upon to supplement system and underground storage supplies.

Peaking needs are met in several ways. Some underground storage sites are designed to provide peaking service, but most often LNG (liquefied natural gas) in storage and liquefied petroleum gas such as propane are vaporized and injected into the natural gas distribution system supply to meet instant requirements. Short-term linepacking is also used to meet anticipated surge requirements.

The use of peaking facilities, as well as underground storage, is essentially a risk-management calculation, known as peak-shaving. The cost of installing these facilities is such that the incremental cost per unit is expensive. However, the cost of a service interruption, as well as the cost to an industrial customer in lost production, may be much higher. In the case of underground storage, a suitable site may not be locally available. The only other alternative might be to build or reserve the needed additional capacity on the pipeline network. Each alternative entails a cost.

A local natural gas distribution company (LDC) relies on supplemental supply sources (underground storage, LNG, and propane) and uses linepacking to "shave" as much of the difference between the total maximum user requirements (on a peak day or shorter period) and the baseload customer requirements (the normal or average) daily usage. Each unit "shaved" represents less demand charges (for reserving pipeline capacity on the trunklines between supply and market areas) that the LDC must pay. The objective is to maintain sufficient local underground natural gas storage capacity and have in place additional supply sources such as LNG and propane air to meet large shifts in daily demand, thereby minimizing capacity reservation costs on the supplying pipeline.

Major Natural Gas Transportation Corridors

Corridors from the Southwest | From Canada | From Rocky Mountain Area | Details about Transportation Corridors

The national natural gas delivery network is intricate and expansive, but most of the major transportation routes can be broadly categorized into 11 distinct corridors or flow patterns.

- 5 major routes extend from the producing areas of the Southwest
- 4 routes enter the United States from Canada
- 2 originate in the Rocky Mountain area.

A summary of the major corridors and links to details about each corridor are provided below.

Major Natural Gas Transportation Corridors

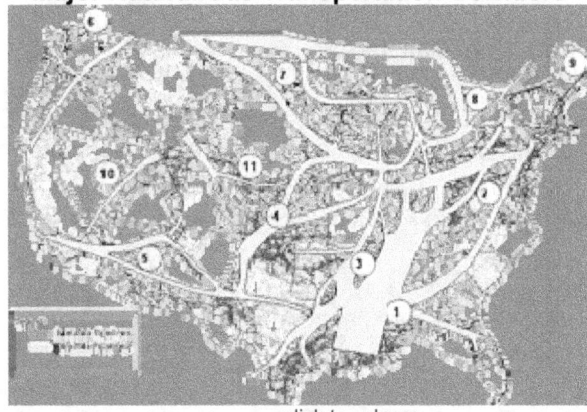

click to enlarge

U.S. Regional Definitions - Map

More information related to Natural Gas Transportation Corridors ...

Region-to-Region System Capacity Levels - Map
Interstate Pipeline Capacity on a State-to-State Level
- spreadsheet

Corridors from the Southwest Region

More than 20 of the major interstate pipelines originate in the Southwest Region. Some extend to the Southeast through Louisiana and Arkansas, others to the Central and Midwestern States through Texas, Oklahoma, and Arkansas, and to the Western States through New Mexico. This area of the country exports about 45 percent (6.0 trillion cubic feet in 2005) of its production, which is 46 percent of the total natural gas consumed elsewhere in the lower 48 States.

Pipelines exiting the region have the capacity to accommodate as much as 40.7 Bcf per day: 58 percent to the Southeast Region, 24 percent to the Central Region, 15 percent to the Western Region, and the rest to Mexico. Much of the pipeline capacity directed toward the Southeast traverses the region en route to Midwestern and Northeastern markets. To a lesser degree, this is also true for the pipeline capacity exiting to the midsection of the country, much of which is ultimately destined for the Midwestern States.

- 1. Southwest-Southeast: from the area of East Texas, Louisiana, and the Gulf of Mexico, to the Southeastern States.

- 2. Southwest-Northeast: from the area of East Texas, Louisiana, and the Gulf of Mexico, to the U.S. Northeast (via the Southeast Region).

- 3. Southwest-Midwest: from the area of East Texas, Louisiana, Gulf of Mexico, and Arkansas to the Midwest.

- 4. Southwest Panhandle-Midwest: from the area of southwestern Texas, the Texas and Oklahoma panhandles, western Arkansas, and southwestern Kansas to the Midwest.

- 5. Southwest-Western: from the area of southwestern Texas (Permian Basin) and northern New Mexico (San Juan Basin) to the Western States, primarily California.

Corridors from Canada

- 6. Canada-Western: from the area of Western Canada to Western markets in the United States, principally California, Oregon, and Washington State.

- 7. Canada-Midwest: from the area of Western Canada to Midwestern markets in the United States.

- 8. Canada-Northeast: from the area of Western Canada to Northeastern markets in the United States.

- 9. Eastern Offshore Canada-Northeast: from the area of offshore eastern Canada (Sable Island) to New England markets in the United States.

Corridors from the Rocky Mountain Area

In the Central Region, only one major interstate pipeline originating within the region provides transportation services directly to another region, Kern River Transmission Company. All the others operate primarily within the Central Region itself or originate in other regions. Shippers using these lines to move supplies outside the region take advantage of the interconnections these lines have with the interstate pipelines traversing the region, principally those coming out of the Southwest Region.

- 10. Rocky Mountains-Western: from the Rocky Mountain area of Utah, Colorado, and Wyoming to the Western States, primarily Nevada and California with support for markets in Oregon and Washington.

- 11. Rocky Mountains-Midwest: from the Rocky Mountain area to the Midwest, including markets in Iowa, Missouri, and eastern Kansas.

Details about the Transportation Corridors

Southwest to Southeastern US

Two fairly distinct subcorridors extend into the Southeast Region from the Southwest: one goes eastward into Mississippi and continues further east, and the second goes northward into Tennessee and Kentucky. Along the first route, there are four major interstate pipeline companies that operate almost exclusively within the Southeast Region -- Florida Gas Transmission Company (FGT), Gulf South Pipeline Company, Gulfstream Natural Gas System, and Southern Natural Gas Company (SONAT). Together they can handle at least 9.6 billion cubic feet (Bcf) per day for shippers in the region.

Varying amounts of capacity on several other large interstate pipelines that follow this subcorridor also serve limited markets in the region. For instance, Transcontinental Gas Pipeline Company (Transco) serves customers in Georgia, South Carolina, and North Carolina as it continues along its route up the east coast. However, this service only represents about 1.1 Bcf per day, or 30 percent, of the 3.5 Bcf per day found on the Transco system as it enters the region. Yet, in North Carolina it is essentially the only source of natural gas supplies to the State.

Along the second subcorridor, one pipeline company Texas Gas Transmission Company (TGT) predominates, at least in terms of delivery points. While this system extends into the Midwest Region, more than 70 percent of its delivery points are located in the States of Kentucky and Tennessee. TGT provides substantial deliveries to underground storage facilities in northern Kentucky that supplement supplies to the local market and to the Midwest Region during the heating season.

Tennessee Gas Pipeline Company (Tenneco) and Texas Eastern Transmission Company (TETCO) are two additional systems operating along this subcorridor, but most of their delivery points are outside the Southeast Region. Tenneco, however, is the principal supplier of gas to two regional interstate pipelines: Enbridge Pipelines (AlaTenn), mostly operating in northern Alabama, and the East Tennessee Gas Company (Tennessee and Virginia).

The underground storage facilities located along this corridor are defined by their location. Those facilities at the corridor's southern end in Louisiana, Mississippi, and Alabama are mainly high-deliverability salt storage sites to support shippers and traders who want to acquire supplies for shipment to market. Of the 11.6 Bcf of daily storage deliverability (withdrawal) available in the area, 56 percent is from salt cavern sites.

This feature provides shippers using these corridors access to very flexible storage, which can be used to enhance their deliverability schedule, avoid transportation imbalances, and support any gas trading or hedging activities they may wish to engage in. In northwestern Kentucky, along the western subcorridor, storage facilities are devoted primarily to providing seasonal supplies. They are supported, for the most part, by deliveries from the Texas Gas Transmission system. The majority of the storage in Mississippi and Alabama is available to shippers using either subcorridor.

Southwest to Northeastern US

The Southwest-to-Northeast corridor consists of two routes. The first extends from East Texas and Louisiana northeastward through Mississippi, Tennessee, Kentucky, and parts of Ohio to enter the Northeast Region via West Virginia or Pennsylvania. The second route begins as the first but then extends northeastward from Mississippi via the east coast States and enters Virginia from the south. The principal interstate pipeline systems operating along the corridor include Tennessee Gas, Columbia Gulf Transmission, and Texas Eastern Transmission on the western segment, and Transcontinental Gas Pipeline on the eastern segment. These four pipeline companies represent approximately 10.2 Bcf per day of total corridor capacity, making this corridor the largest of the major transportation corridors in North America.

During wintertime peak periods, each of the systems is almost fully utilized. During the summer months, however, usage rates for the pipeline systems operating along this corridor tend to drop substantially. The principal factor affecting summertime usage rates on several of these pipeline systems is the demand for gas to refill underground storage sites in the States of West Virginia and Pennsylvania, and, to some degree, Ohio and New York as well.

The majority of the more than 190 underground storage sites located along this corridor are accessible to shippers. At the southwestern terminus of the corridor, more than 30 sites with a working gas capacity of at least 624 billion cubic feet and a daily withdrawal capability of 13 Bcf per day are located within 20 miles of the subject pipeline systems. Most of this capacity is used by producers, who use it to store short-term excess production, and by market centers.

This corridor links with some of the most active trading points located outside the Southwestern production area. One of the most significant is the Ellisburg-Leidy center in Pennsylvania, which provides interconnections and transportation services between the pipelines comprising this corridor and the other major interstate pipelines operating primarily within the Northeast States. Shippers using the corridor may also utilize the services of several natural gas market centers to expand their marketing and transportation options.

Southwest to Midwestern US

The Southwest-to-Midwest corridor extends northward out of East Texas, Louisiana, and Arkansas (Arkoma Basin production) and generally through Tennessee/Kentucky into the Midwest Region, although a part of it also travels through Missouri. The principal interstate pipeline systems operating along this corridor are: ANR Pipeline Company (ANR), Midwestern Gas Transmission Company (via Tennessee Gas Pipeline Company), Natural Gas Pipeline Company of America (NGPL), Texas Gas Transmission Company (TGT), Texas Eastern Transmission Company (TETCO), and Trunkline Gas Company. Mississippi River Gas Transmission Pipeline Company also transports gas along this corridor but it terminates in the St Louis, Missouri, area. Its operations in Illinois are confined to the area east of St Louis.

The portions of these systems located along this corridor represent approximately 7.8 Bcf per day, or 28 percent of the total pipeline capacity feeding into the Midwest Region (27.5 Bcf per day). They also account for more than 30 percent of the total pipeline capacity exiting this area of the Southwest.

Little underground storage is located along the midsection of this corridor. However, shippers have access to significant amounts of storage at either end. This corridor also links together two major gas trading centers: the Henry Hub in Louisiana and the Chicago Center in northern Illinois. In addition, the corridor also includes several natural gas trading (and price discovery) locations accessible to shippers and traders via the several major commercial electronic trading systems set up in the United States and Canada.

During the heating season, these markets are actively used by shippers and other market participants as a way to balance their receipts/deliveries, for arbitrage between the two markets, and to smooth market and price fluctuations through hedging.

Southwest Panhandle to Midwestern US

This route is a major link between the Waha area (Permian Basin) of southwestern Texas and the Chicago area market. It extends from the West Texas and Oklahoma Panhandle areas northward through the major gas production fields (including Hugoton and Panhandle) located in southwestern Kansas, and then northeastward toward the Midwest marketplace. In Nebraska, it links with another corridor (see Rocky Mountain-Midwest section) bringing supplies in from the Rocky Mountain areas of Wyoming, Utah, and Colorado.

There are four major interstate pipelines that run along this corridor: ANR Pipeline Company, Panhandle Eastern Pipeline Company, Northern Natural Gas Company, and Natural Gas Pipeline Company of America. These four pipelines alone constitute 67 percent of total pipeline capacity exiting this area. These pipeline routes, however, represent only about 17 percent of the total capacity into the Midwest Region. The Trailblazer Pipeline system ties in Rocky Mountain supplies with an interconnection to Natural Gas Pipeline Company of America in Nebraska.

Market centers and commercial trading points located in the Waha and Panhandle area of West Texas serve this transportation corridor at its apex. At its terminus, shippers and traders can link their Texas trading with the Chicago market. In addition, trading centers located in south central Kansas provide shippers with the opportunity to do business with traders in the other two areas. All four pipelines operating in the corridor have direct or indirect links with each other.

Only a limited amount of underground storage capacity is available to transporters along this route. However, during the nonheating season a sizeable amount of capacity on these systems is used to transport supplies for injection into storage facilities in Illinois, Indiana, and Michigan. The ANR Pipeline system in particular has a number of open-access sites located at the northern end of its system in Michigan. NGPL has a number of storage sites located in Illinois.

Southwest to Western US

The Southwest-Western corridor is used to transport supplies from the Permian Basin area of West Texas, through New Mexico (where the northern route taps into the San Juan Basin production area), and westward primarily to Arizona and California. Two major interstate pipelines, El Paso Natural Gas Company and Transwestern Pipeline Company, operate along this corridor. Both of these pipelines end at the California or Nevada State borders, where they deliver supplies to Southwest Gas Company (Nevada), Southern California Gas Company, and Pacific Gas & Electric Company, the largest pipelines serving the California marketplace. In addition, Transwestern Gas Pipeline Company links with the Mojave Pipeline Company, an interstate pipeline that transports natural gas supplies to the enhanced oil recovery (EOR) and cogeneration customers located in Kern County, California.

Joining El Paso Natural Gas Company and Transwestern Pipeline Company along the northern route, in 2002 Questar's Southern Trails Pipeline (an oil pipeline conversion) was completed, transporting an additional 90 MMcf per day between the San Juan Basin area and the California border.

Much of the natural gas flowing along this corridor is produced in the San Juan Basin. The TransColorado Pipeline system, completed in 1996, can move as much as 590 MMcf per day from north central Colorado and the Ignacio area of the southern Colorado San Juan Basin to interconnections with the El Paso Natural Gas and Transwestern Pipeline systems in the Blanco area of northwestern New Mexico. Northwest Pipeline Company also can deliver up to 250 Mmcf per day into these two systems.

A significant amount of West Texas and New Mexico gas supplies also are transported along the southern portion of this corridor, which consists primarily of the El Paso Natural Gas Company's Line 2000 which has a throughput capacity of approximately 2.4 Bcf per day. This section of the corridor primarily serves southern Arizona and southern California, but in 2002 it also began service to the new North Baja Pipeline system, designed to transport up to 500 MMcf per day to Mexico. (In 2010, the North Baja Pipeline system will become bidirectional, having the capability to transport up to 2.0 Bcf per day from LNG import facilities in Mexico to Western U.S. markets.)

There is very little underground natural gas storage capacity associated with this corridor. At the extreme eastern end of the corridor, only one site, the Washington Ranch facility operated by El Paso Natural Gas Company, is reserved primarily for system support services and is not available for customer use. At its western end, in southern California, a limited amount of storage capacity is available to shippers at five sites operated by Southern California Gas Company (SoCal).

Although some of the natural gas injected into these storage sites comes from producing fields in southern California, a significant amount of the working gas stored at these sites comes out of this corridor. The combined withdrawal rate capability of the four sites is 3.7 billion cubic feet (Bcf) per day, while their total working gas capacity is 120 Bcf. This translates into roughly 32 days of backup from these sites.

Western Canada to Midwestern US

This transportation corridor lies between Western Canadian supply areas and the U.S. Midwest and links two Canadian systems, TransCanada Pipeline Ltd. and Foothills Pipeline Company, with three United States pipeline systems, Great Lakes Gas Transmission Company, Northern Border Pipeline Company and Viking Gas Transmission Company. In addition, the 1,300-mile Alliance Pipeline, completed in late 2000, provides a direct transportation route for "wet" (natural gas high in liquids content) between producing fields in northwestern British Columbia and Alberta, Canada, and a gas-processing plant (Aux Sable) located outside Chicago, Illinois. These tie-ins represent about 6.2 Bcf per day of pipeline capacity, or about 41 percent of total U.S. natural gas import capacity in 2006. Between 1990 and 2006, capacity on this route more than doubled, going from 3.1 Bcf per day in 1990 to 7.2 Bcf per day in 2006.

The Northern Border Pipeline (NBP) system extended its pipeline system to Illinois (from it original terminus in Iowa) in 1998 and to Indiana in 2001, now providing almost a Bcf per day to the Chicago area and to customers in Indiana. In 2000, another pipeline, the Vector Pipeline system, located between Chicago, Illinois and Dawn, Ontario, at the eastern end of the corridor, was placed in service. It can transport up to one Bcf per day between the United States (Michigan) and Canada (Ontario). It was developed primarily to provide an alternative expansion route for Canadian gas and service to customers in Ontario, Canada. It also lies along a route that can be expanded to potentially accommodate gas transportation of Western Canadian gas to Northeast U.S. markets via the Empire/Millennium gas pipeline system that is scheduled for development in New York State in 2008.

A large number of underground storage facilities are located in proximity to several of the pipeline systems operating in this corridor, although not all of them are directly accessible to shippers. For instance, nine sites (1 Bcf per day injection, 1.8 Bcf per day withdrawal capability) are directly accessible to shippers using the Great Lakes Gas Transmission system, while the storage facilities located in Illinois and operated by Northern Illinois Gas Company (eight sites, 3.4 Bcf working gas capacity) are available only through the Chicago Market Center, which is affiliated with the company, or through the company itself. Altogether, the daily injection capability at storage facilities linked to the receiving end of this corridor represents the potential use of about five Bcf per day of pipeline capacity during the storage refill period from April through October.

Western Canada to Northeastern US

The western portion of the Canada-Northeast corridor links the TransCanada Pipeline system (and Western Canadian gas production) to seven pipeline companies in the Northeastern United States. The seven are: Iroquois Pipeline Company, North Country Pipeline Company, the Portland Gas Transmission System, Tennessee Gas Pipeline Company, Empire Pipeline Company, Vermont Gas Company, and St. Lawrence Gas Company. Indirectly, the corridor also supplies gas to the National Fuel Gas Supply Company and Dominion Transmission Company.

The seven systems transport gas primarily into New York and the New England States at a total capacity level of 3.4 Bcf per day. While the vast majority of the Canadian capacity that comes into the U.S. Northeast is off the northern tier of the TransCanada system, about five percent represents capacity that traverses the U.S. Midwest (on the Great Lakes Transmission system), crosses back into Canada through Ontario, and is imported once again at Niagara, New York.

In Canada, at the western end of this corridor in Alberta and Saskatchewan provinces, approximately 4 Bcf per day of daily storage deliverability is available at 12 sites interconnected with the TransCanada Pipeline System. In addition, over 25 storage sites located at Dawn, Ontario, Canada, are available to shippers transporting supplies to the area via the Great Lakes Transmission system. In the U.S. Northeast, storage deliverability of up to 14.8 Bcf per day is available to these shippers.

Eastern Canada to Northeast (New England)

This corridor consists primarily of the Maritimes and Northeast Pipeline system, completed in late 1999. It can transport more than 445 MMcf per day into the United States from off the eastern coast of Canada at Sable Island. The current system merges with the Portland Gas Transmission System at Wells, Maine to deliver almost 628 MMcf per day in northern Massachusetts to customers on the Tennessee Gas Pipeline system. Beginning in 2002, with the completion of Phase III of the Maritimes and Northeast Pipeline system, shippers have had the option of transporting up to 230 MMcf per day of this capacity to the Boston, Massachusetts area on the system.

Canada to Western US

The Canada-Western route brings natural gas from Alberta and British Columbia, Canada, through the States of Washington, Idaho, and Oregon, with terminating points in Nevada and California. In Canada,

Spectra Energy Corporation's Westcoast Gas Transmission Ltd. and Alberta Natural Gas Ltd. (in association with Foothills Pipeline Ltd.) receive gas from the TransCanada Pipeline (NOVA) in Alberta (the principal pipeline system in the region linked into the major production areas in Alberta and British Columbia) and transport that gas to the U.S. border. There the supplies are received by Northwest Pipeline Company (from Westcoast Gas Transmission) and Gas Transmission Northwest from Alberta Natural Gas. The two pipelines have a combined capacity of 4.4 Bcf per day, 99 percent of import capacity in the area. This route represents one-quarter of the total capacity reaching the United States from Canada.

While the Gas Transmission Northwest Company transports most of its gas, about 76 percent in 2006, directly southward to California, the Northwest Pipeline Company system extends south and eastward from its border receipt point, operating on a bidirectional basis along much of the eastern section. At the northern Nevada State line, Northwest Pipeline Company links with the Paiute Pipeline Company, which until recently was the only gas supplier to the Reno, Nevada, area. Only one new pipeline has been added to the corridor since 1990, the Tuscarora Pipeline Company (110 MMcf per day) in 1995. This pipeline interconnects with the Gas Transmission Northwest Company system at the northern California border and transports gas to the Reno, Nevada area.

Access to underground storage for shippers along this corridor is limited. Much of the storage capacity on the southern portion is owned and operated by local distribution companies and is used exclusively to support their own seasonal storage needs. Nevertheless, shippers can acquire access to storage services on an as-available basis through several independent storage operations. The California Gas Transmission Company provides limited access to its three storage sites in northern California. At the Canadian end of the corridor, much of the available storage is intricately linked with market center operations, providing parking and loaning services primarily to producers shipping gas to the United States. These Canadian sites are capable of handling up to 6 Bcf per day deliverability and have a working gas capacity level of about 412 Bcf.

Rocky Mountain Area to Western US

This system extends from the Opal, Wyoming area southwestward through Nevada, just north of Las Vegas, to Kern County, California. In California, the Kern River Pipeline system physically merges with the Mojave Pipeline system (400 MMcf per day) to form one line serving customers primarily in Kern and San Bernardino Counties in California. Mojave receives its supplies from Transwestern Gas Pipeline Company and El Paso Natural Gas Company at the Arizona-California border. Its capacity is approximately 885 million cubic feet per day.

The Kern River Pipeline system was developed primarily to carry gas to the enhanced oil recovery market in southern California, which has been a large natural gas market. In 1997, its service was extended to the Las Vegas electric power generation market with the opening of an expanded metering facility with Southwest Gas Company, the major natural gas distributor in the Las Vegas area. Its system capacity was doubled in 2003 to approximately 1,750 MMcf/d to accommodate the growing demand along its route.

Underground storage facilities, although available at the apex of this corridor in Wyoming and Utah, do not play a major role in the operations of the Kern River Pipeline system. Although six sites are in the vicinity, with a combined daily deliverability of 0.7 Bcf per day and 57 Bcf of working gas capacity, only one, Questar Pipeline Company's Clay Basin facility (0.4 Bcf per day, 51 Bcf), is accessible to shippers.

Rocky Mountain Area to Midwestern US

This corridor links Rocky Mountain natural gas supplies from Utah, Wyoming, and Colorado with markets in the Midwestern United States and with several sizable metropolitan markets in eastern Kansas and Missouri. While the corridor itself does not yet extend into the Midwest, the several pipelines currently operating along this route interconnect with major trunklines that bring natural gas supplies from the Southwest Region to Midwestern markets.

The Trailblazer System, which is a contiguous linkup of the Overthrust, Wyoming Interstate, and Trailblazer pipelines, operates from western Wyoming to eastern Nebraska, where it offloads to the Natural Gas Pipeline Company of America pipeline. Similarly, Colorado Interstate Pipeline Company 's Cheyenne Plains Pipeline, built in 2004, provides more than 730 MMcf/d of gas transportation for Wyoming and Colorado production from the Cheyenne Hub located in northeastern Colorado. The Cheyenne Plains Pipeline terminates with interconnections to Northern Natural Gas and Natural Gas Pipeline Company of America in southwestern Kansas. Natural gas transported on these pipeline systems is subsequently delivered to customers in the eastern portion of the Central Region and in Midwestern markets.

The Southern Star Central Pipeline (formerly Williams Natural Gas Co – Central) and the KM Interstate Pipeline Company also have operations along this corridor, but these two pipelines serve primarily local regional markets. However, the KM Interstate Pipeline Company system does include its Pony Express Pipeline (255 million cubic feet per day) segment which runs from central Wyoming to south of Kansas City, Missouri. Currently this segment does not provide any interconnections with the two major interstate pipelines connecting this corridor to Midwestern markets; rather, its full capacity is committed to customers located in the Kansas City area.

In 2008, the first segment of the Rockies Express Pipeline system, which is the second segment of a 1,663-mile, 1.8 Bcf per day, pipeline project that would transport Rocky Mountain natural gas to Midwest and eventually to Northeast markets, is expected to be placed in service. The first segment of the new system, completed in early 2007, involved the construction of a 327-mile pipeline system from the Meeker Hub in Rio Blanco County, Colorado, to the Cheyenne Hub in Weld County, in northeastern Colorado. Completion of the entire system, which is scheduled for early 2010, will mark the first time that Rocky Mountain natural gas supplies would be delivered directly to Midwest markets.

Customers using this corridor have a limited number of underground storage facilities available for their use. At the terminus of the corridor in Wyoming and Colorado are 18 sites that customers may access. Much of the storage located at this end, however, is used to support local producers and distribution companies. In the Chicago area corridor, shippers also have access to several storage facilities associated with the Chicago market center.

About U.S. Natural Gas Pipelines - Transporting Natural Gas

Underground Natural Gas Storage

Overview | Regional Breakdowns

Overview

Underground natural gas storage provides pipelines, local distribution companies, producers, and pipeline shippers with an inventory management tool, seasonal supply backup, and access to natural gas needed to avoid imbalances between receipts and deliveries on a pipeline network.

There are three principal types of underground storage sites used in the United States today. They are:

- depleted natural gas or oil fields,
- aquifers, or
- salt caverns.

In a few cases mine caverns have been used. Most underground storage facilities, 81 percent at the beginning of 2006, were created from reservoirs located in depleted natural gas production fields that were relatively easy to convert to storage service, and that were often close to consumption centers and existing natural gas pipeline systems.

U.S. Underground Natural Gas Storage Facilities

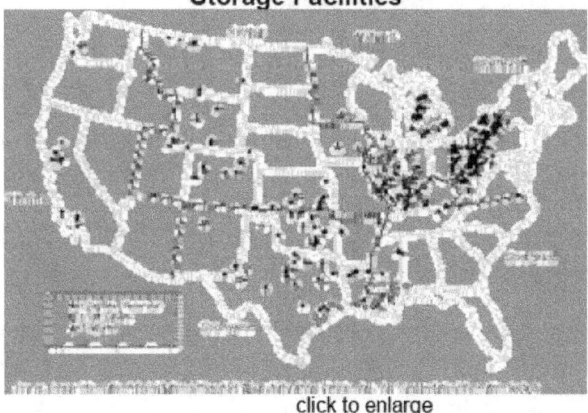

click to enlarge

More information related to underground storage...

Depleted Reservoir Storage Well - illustration
Aquifers - illustration
Salt Caverns - illustration

Underground Storage Capabilities by State – Appendix table

Underground Storage Field Level Data from Form EIA-191A - spreadsheet

In some regions, such as the Midwestern United States, suitable natural aquifers have also been converted for use as natural gas storage facilities. An aquifer is usable for natural gas storage if the water-bearing sedimentary rock formation is overlaid with an impermeable cap rock. While the geology of aquifers is similar to that of depleted production fields, their use in gas storage usually requires more base (cushion) gas and greater monitoring of withdrawal and injection performance.

Since the 1980s, the number of salt cavern storage sites developed in the United States has grown steadily, principally because of its unique capabilities and high cycling rate (inventory turnover). The large majority of salt cavern storage facilities have been developed in salt dome formations located in the Gulf Coast States. Salt caverns leached from bedded salt formations in Northeastern, Midwestern, and Western States have also been developed but the number has been limited due to a lack of suitable geology. Cavern construction is more costly than depleted field conversions when measured on the basis of dollars per thousand cubic feet of working gas capacity, but the ability to perform several withdrawal and injection cycles each year reduces the per-unit cost of each thousand cubic feet of gas injected and withdrawn.

Underground Storage by U.S. Region

At the close of 2006, 398 underground natural gas storage sites were operational in the United States. During the year, four new storage sites were added, one in Colorado, New York, Texas, and California, while 13 existing storage fields underwent expansions. Consequently, working gas capacity in the U.S. increased by 49 Bcf, to 4,059 Bcf (4,010 Bcf in 2005) while deliverability rates rose to 85.1 Bcf/d (84.2 Bcf/d in 2005). The largest expansion of working gas capacity (13 Bcf) occurred at the Stagecoach natural gas storage site in New York State, a depleted-reservoir facility. Depleted-reservoir storage accounted for about 92 percent of the 49 Bcf of new working gas capacity added in 2006.

The number, type, and profile of underground natural gas storage varies by region. Below is a brief overview for each of the six regions in the lower 48 States.

. Central (50) | Midwest (121) | Northeast (108) | Southeast (33) | Southwest (67) | Western (19) | Overall (398)

Central Region

Underground natural gas storage in the Central Region is notable for several reasons. First, many of the 49 storage facilities located in the region are used to store excess production rather than to serve as a supply source for local markets. Production is stored to allow a stable flow rate despite temporary demand fluctuations.

Second, the region has the Nation's largest storage site, the Baker/Cedar Creek Field in Montana, with a total capacity of 287 billion cubic feet (Bcf) and a working gas capacity of 164 Bcf. The total regional working gas storage capacity (approximately 556 billion cubic feet) is 14 percent of the U.S. total, while daily deliverability from storage is only 6.1 billion cubic feet per day, or 7 percent of the U.S. total.

Central Region Summary of Underground Natural Gas Storage, by State & Reservoir Type, 2006

State	Depleted Gas Oil Fields			Aquifer Storage			Salt Cavern Storage			Total		
	Sites	Working Gas Capacity (Bcf)	Daily Withdrawal Capability (MMcf)	Sites	Working Gas Capacity (Bcf)	Daily Withdrawal Capability (MMcf)	Sites	Working Gas Capacity (Bcf)	Daily Withdrawal Capability (MMcf)	Sites	Working Gas Capacity (Bcf)	Daily Withdrawal Capability (MMcf)
Colorado	9	42	1,088	0	0	0	0	0	0	9	42	1,088
Iowa	0	0	0	4	75	1,025	0	0	0	4	75	1,025
Kansas	18	117	2,348	0	0	0	1	1	0.1	19	118	2,348
Missouri	0	0	0	1	11	350	0	0	0	1	11	350
Montana	5	196	300	0	0	0	0	0	0	5	196	300
Nebraska	1	16	169	0	0	0	0	0	0	1	16	169
North Dakota	*0*	*0*	*0*	*0*	*0*	*0*	*0*	*0*	*0*	*0*	*0*	*0*
South Dakota	*0*	*0*	*0*	*0*	*0*	*0*	*0*	*0*	*0*	*0*	*0*	*0*
Utah	1	51	427	2	1	100	0	0	0	3	52	527
Wyoming	7	42	227	1	4	75	0	0	0	8	46	302
Subtotal	41	464	4,559	8	91	1,550	1	1	0	50	556	6,109
(Marginal Sites)[1]	(7)	(4)	(145)	(0)	(0)	(0)	(1)	(1)	(0)	(8)	(5)	(145)
Percent of U.S.	13	13	7	18	23	19	3	1	0	13	14	7

[1] Marginal sites: Very little or no activity reported during the 2006 calendar year. Marginal sites included in State/Regional totals.
Note: Bcf = Billion cubic feet. MMcf = Million cubic feet. States with no undergrond natural gas storage are shown in *Italics*.
Source: Energy Information Administration, GasTran Natural Gas Transportation Information System, Underground Natural Gas Storage Database.

The Baker/Cedar Creek Field, owned and operated by the Williston Basin Interstate Pipeline Company, serves as support infrastructure for the natural gas that is produced in association with oil production in the area. With an estimated peak-day withdrawal rate of about 134 million cubic feet per day, the flow from this storage field is directed primarily to interconnections with the Northern Border Pipeline Company system between North and South Dakota. In recent years, however, the Baker field has not been heavily utilized due to a decrease in production from nearby associated-gas fields.

Storage facilities in Kansas, specifically in the southeastern portion of the region, provide service to the interstate pipeline systems that move natural gas to the Midwest Region, but they are also integral to regional requirements. For instance, about 35 percent of the State's working gas storage capacity of approximately 118 billion cubic feet is owned and operated by Southern Star Central Gas Pipeline Company, which is primarily a regional interstate pipeline.

About 96 percent of the storage capacity in Kansas is available to customers and shippers on other interstate trunklines, while the remaining 4 percent is devoted to local distribution and production field service. About 40 percent of the daily peak-day storage deliverability in the State, or 940 million cubic feet per day, is available to the two interstate pipeline companies traversing the region, Northern Natural Gas Company and Panhandle Eastern Pipe Line Company.

Storage facilities in the rest of the region serve primarily as seasonal supply sources for local markets. Storage fields in Utah provide service to shippers on the Questar Pipeline Company system as well as to customers within the Salt Lake City area. The storage fields in Colorado and portions of Wyoming service the Denver area through the Colorado Interstate Gas Company system. The local distribution companies serving these markets account for about 16 percent of the total storage deliverability in the region

Midwest Region

Many of the pipelines serving the region also provide their shipper/customers with access to a large amount of underground storage capacity located in Illinois, Indiana, Michigan, and Ohio. The Midwest Region has 121 sites, the largest number in the country.

Of all six regions, this region has the largest volume of underground (working gas) storage capacity (almost 1.2 trillion cubic feet (Tcf)) and daily deliverability (26.4 Bcf/d) from storage. These levels account for about 30 percent of the U.S. total in each category. Regional intrastate pipelines and/or local distribution companies, such as Northern Illinois Gas Company (NICOR), control about 61 percent of daily deliverability from storage in this region.

Midwest Region Summary of Underground Natural Gas Storage, by State & Reservoir Type, 2006

State	Depleted Gas/Oil Fields			Aquifer Storage			Salt Cavern Storage			Total		
	Sites	Working Gas Capacity (Bcf)	Daily Withdrawal Capability (MMcf)	Sites	Working Gas Capacity (Bcf)	Daily Withdrawal Capability (MMcf)	Sites	Working Gas Capacity (Bcf)	Daily Withdrawal Capability (MMcf)	Sites	Working Gas Capacity (Bcf)	Daily Withdrawal Capability (MMcf)
Illinois	11	51	835	18	256	5,294	0	0	0	29	307	6,129
Indiana	10	14	261	12	20	501	0	0	0	22	34	762
Michigan	43	632	14,636	0	0	0	2	2	85	45	634	14,721
Minnesota	0	0	0	1	2	60	0	0	0	1	2	60
Ohio	24	220	4,692	0	0	0	0	0	0	24	220	4,692
Wisconsin	0	0	0	0	0	0	0	0	0	0	0	0
Active Sites	88	917	20,424	31	278	5,855	2	2	85	121	1,197	26,364
(Marginal Sites)[1]	(8)	(8)	(119)	(1)	(2)	(43)	(0)	(0)	(0)	(9)	(10)	(162)
Percent of U.S.	27	26	33	70	70	70	6	1	1	30	29	31

[1] Marginal sites: Very little or no activity reported during the 2006 calendar year. Marginal sites included in State/Regional totals.

Note: Bcf = Billion cubic feet. MMcf = Million cubic feet. States with no undergrond natural gas storage are shown in *italics*.

Source: Energy Information Administration, GasTran Natural Gas Transportation Information System, Underground Natural Gas Storage Database.

In Illinois, 50 percent of the daily deliverability from storage is integrated into three pipeline or distribution systems: Northern Illinois Gas Company, Illinois Power Company, and Central Illinois Public Service Company. Northern Illinois Gas Company also provides access to part of its working gas storage to support shippers using the regional Chicago natural gas market center.

The Great Lakes Gas Transmission Company and the ANR Pipeline Company systems both use Michigan storage facilities extensively to support their shippers' needs. In the first case, the Great Lakes Transmission Company system transports most of its volume eventually to markets in Ontario, Canada, but it uses storage sites located in Michigan to store supplies shipped for Canadian customers during the summer, providing withdrawal and delivery services during winter peak periods. ANR Pipeline Company and its affiliate ANR Storage Company together operate 13 sites in the State, while other storage operators in the State include the MichCon Gas Company (four sites), the Michigan Gas Storage Company (three sites), and its parent Consumers Energy Company (11 sites).

Elsewhere in the Midwest Region, Consumers Energy Company, with 14 sites in Michigan, is the single largest LDC operator of underground storage fields in the lower 48 States. Its sites have an overall deliverability of more than 4.0 Bcf/d and working capacity of 154 Bcf. Trailing closely is the Northern Illinois Gas Company, which operates eight natural gas storage facilities in Illinois with a total daily deliverability level of 3.1 Bcf/d and a total working gas capacity level of 152 Bcf.

Northeast Region

The States of Pennsylvania and New York are the key transit areas for gas deliveries within the region and include the major service territories of Dominion Transmission Company and Columbia Gas Transmission Company systems. These States, along with West Virginia, also have the largest underground storage capacity in the region. Storage is essential as a supply backup and for balancing gas supplies on the pipelines operating in the region. More pipeline capacity exits these States than enters, reflecting their storage capability as a seasonal supply source for the States to the north and east.

Northeast Region Summary of Underground Natural Gas Storage, by State & Reservoir Type, 2006

State	Depleted Gas Oil Fields			Aquifer Storage			Salt Cavern Storage			Total		
	Sites	Working Gas Capacity (Bcf)	Daily Withdrawal Capability (MMcf)	Sites	Working Gas Capacity (Bcf)	Daily Withdrawal Capability (MMcf)	Sites	Working Gas Capacity (Bcf)	Daily Withdrawal Capability (MMcf)	Sites	Working Gas Capacity (Bcf)	Daily Withdrawal Capability (MMcf)
Connecticut	0	0	0	0	0	0	0	0	0	0	0	0
Delaware	0	0	0	0	0	0	0	0	0	0	0	0
Maine	0	0	0	0	0	0	0	0	0	0	0	0
Maryland	1	17	400	0	0	0	0	0	0	1	17	400
Massachusetts	0	0	0	0	0	0	0	0	0	0	0	0
New Hampshire	0	0	0	0	0	0	0	0	0	0	0	0
New Jersey	0	0	0	0	0	0	0	0	0	0	0	0
New York	23	116	1,840	0	0	0	1	2	145	24	118	1,985
Pennsylvania	49	405	8,635	0	0	0	0	0	0	49	405	8,635
Rhode Island	0	0	0	0	0	0	0	0	0	0	0	0
Vermont	0	0	0	0	0	0	0	0	0	0	0	0
Virginia	1	1	22	0	0	0	2	4	325	3	5	347
West Virginia	31	244	3,701	0	0	0	0	0	0	31	244	3,701
Active Sites	105	783	14,598	0	0	0	3	6	470	108	789	15,068
(Marginal Sites)[1]	(6)	(29)	(32)	(0)	(0)	(0)	(0)	(0)	(0)	(6)	(29)	(32)
Percent of U.S.	33	23	23	0	0	0	10	3	3	27	19	18

[1] Marginal sites: Very little or no activity reported during the 2006 calendar year. Marginal sites included in State/Regional totals.

Note: Bcf = Billion cubic feet. MMcf = Million cubic feet. States with no undergrond natural gas storage are shown in *Italics*.

Source: Energy Information Administration, GasTran Natural Gas Transportation Information System, Underground Natural Gas Storage Database.

The largest storage operators in the Northeast are also three of the largest pipeline companies in the region. Columbia Gas Transmission Company operates 29 storage facilities (out of 107 within the region), with a working gas storage capacity of 140 Bcf (out of a total 767 Bcf). Although Dominion Transmission Company operates only 14, its facilities have the largest amount of working gas capacity in the region,

409 Bcf. National Fuel Gas Supply Company operates the largest number of storage facilities in the region, 31, but its storage fields are only capable of storing up to 114 Bcf of working gas.

(Note: The peak-day deliverability from LNG in the region, 3.8 Bcf per day, is 26 percent as large as the total daily deliverability from underground storage facilities. This backup capability is incorporated into the operations of the regional network and is used to meet the rapid increases in demand that can occur because of sudden temperature changes in the region. Two of the five currently active LNG importing facilities in the U.S. are located in the Northeast Region, the Cove Point LNG Company facility, located on the eastern shore of Maryland, and the DistriGas Company's Everett LNG facility located outside of Boston, Massachusetts.)

Southeast Region

The Texas Gas Transmission Company is the only long-haul natural gas pipeline system that retains a large portion of its deliverability for regional service, primarily in Kentucky. Slightly over 50 percent of its deliverability is within the region. This pipeline company also provides its shippers access to five company-owned underground storage facilities with a working gas capacity of more than 80 Bcf, or about 66 percent of the total working gas capacity in the State of Kentucky (and 45 percent of the regional total).

That level of storage service is equivalent to about 80 percent of the total daily capacity of Texas Gas Transmission Company's lines moving north into Indiana. These storage facilities are also in close proximity to the ANR Pipeline Company system, which traverses the State of Kentucky to Indiana and Michigan. Combined, Texas Gas Transmission Company and ANR Pipeline Company have the capacity to move 2.9 Bcf per day north into Indiana.

In addition to the conventional underground storage facilities located in the northern portion of the region, a number of high-deliverability (salt cavern) storage sites have been built during the past decade in the southern portion to better serve a restructured U.S. natural gas pipeline industry. Four such sites are now located in Mississippi (three) and Alabama (one), with several more planned. The availability of these sites has made these two States a prime market for the type of storage services needed by shippers with high upstream demand swings and local load balancing requirements. These sites are used by customers and shippers doing business on Florida Gas Transmission Company, Gulf South Pipeline Company, Tennessee Gas Pipeline Company, Transcontinental Gas Pipeline Company, and Southern Natural Pipeline Company systems.

Southwest

Underground natural gas storage plays a vital role in the efficient export and transmission of natural gas from the Southwest to other regions, as well as in supplementing regional needs. Its 67 underground storage facilities represent 1,036 Bcf of working gas capacity and an estimated daily deliverability level of 23.1 million cubic feet.

Only the Midwest Region has more working gas capacity and daily deliverability from storage. A large portion of regional storage is near production fields and is used to balance production flows with fluctuating market demand.

In recent years, however, an increasing amount of the area's storage capacity is being developed and used to support regional natural gas market center/hub operations. It is also being used as high-deliverability storage (from salt-caverns) to serve the growing number of variable-load customers, such as gas-fired power plants, that are emerging in the region and which have a need for rapid access to stored natural gas working gas.

Southwest Region Summary of Underground Natural Gas Storage, by State & Reservoir Type, 2006

State	Depleted Gas Oil Fields			Aquifer Storage			Salt Cavern Storage			Total		
	Sites	Working Gas Capacity (Bcf)	Daily Withdrawal Capability (MMcf)	Sites	Working Gas Capacity (Bcf)	Daily Withdrawal Capability (MMcf)	Sites	Working Gas Capacity (Bcf)	Daily Withdrawal Capability (MMcf)	Sites	Working Gas Capacity (Bcf)	Daily Withdrawal Capability (MMcf)
Arkansas	2	15	231	0	0	0	0	0	0	2	15	231
Louisiana	8	286	3,999	0	0	0	6	47	3,153	14	333	7,152
New Mexico	2	54	310	1	2	3	0	0	0	3	56	313
Oklahoma	13	188	3,870	0	0	0	0	0	0	13	188	3,870
Texas	20	355	4,840	0	0	0	15	89	6,646	35	444	11,486
Active Sites	45	898	13,250	1	2	3	21	136	9,799	67	1,036	23,052
(Marginal Sites)[1]	(6)	(13)	(177)	(0)	(0)	(0)	(1)	(4)	(0)	(7)	(17)	(177)
Percent of U.S.	14	26	21	2	1	0	68	74	70	17	26	27

[1] Marginal sites: Very little or no activity reported during the 2006 calandar year. Marginal sites included in State/Regional totals.
Note: Bcf = Billion cubic feet. MMcf = Million cubic feet.
Source: Energy Information Administration, GasTran Natural Gas Transportation Information System, Underground Natural Gas Storage Database.

The States of Louisiana and Texas have more salt-cavern natural gas storage facilities (21) than anywhere else in the United States, of which almost half are owned/operated by independent storage operators. In fact, one-fifth of the working gas capacity and one-third of the daily deliverability available in the region is operated by independents.

About 35 percent of the region's daily storage deliverability remains with interstate pipeline companies, while the rest is operated by LDCs (33 percent) or intrastate pipeline companies (32 percent).

All of the interstate pipeline-owned storage and most of the independently-owned storage is open access, that is, working gas storage capacity that is available to shippers/customers on a first-come, first-served basis at nondiscriminatory rates. The remainder is reserved for system or pipeline use, such as load balancing operations.

While only about a third of the region's storage capacity is owned by LDCs and used exclusively for local service, regional distributors also have access to and use interstate and independent storage facilities. Most of the LDC-owned storage is near major industrial and population centers and has little impact upon the interstate pipeline network in the area. In Texas and Oklahoma, approximately 40 percent of underground storage capacity is at facilities operated by LDCs or intrastate pipeline companies, whereas in Arkansas all of the storage capacity is controlled by local operators.

Western Region

Underground natural gas storage facilities are found in only half of the states in the region, California, Oregon, and Washington. Moreover, approximately 88 percent of the region's working gas capacity is located in California's 12 underground natural gas storage sites, all but two of which are owned by the two principal gas distributors in the State, Southern California Gas Company (SoCal) and Pacific Gas and Electric Company (PG&E). Most of their storage capacity is used for system balancing and as a way of maintaining a steady and high-utilization of pipeline capacity directed from Canada and the Southwest.

The two independent storage facilities in California are used primarily as depositories for gas produced within the State that is not immediately marketable. In addition, these sites are connected to, and deliver their withdrawals to, the Southern California Gas Company and/or Pacific Gas and Electric Company systems.

Western Region Summary of Underground Natural Gas Storage, by State & Reservoir Type, 2006

State	Depleted Gas/Oil Fields			Aquifer Storage			Salt Cavern Storage			Total		
	Sites	Working Gas Capacity (Bcf)	Daily Withdrawal Capability (MMcf)	Sites	Working Gas Capacity (Bcf)	Daily Withdrawal Capability (MMcf)	Sites	Working Gas Capacity (Bcf)	Daily Withdrawal Capability (MMcf)	Sites	Working Gas Capacity (Bcf)	Daily Withdrawal Capability (MMcf)
Arizona	0	0	0	0	0	0	0	0	0	0	0	0
California	12	272	6,380	0	0	0	0	0	0	12	272	6,380
Idaho	0	0	0	0	0	0	0	0	0	0	0	0
Nevada	0	0	0	0	0	0	0	0	0	0	0	0
Oregon	6	14	493	0	0	0	0	0	0	6	14	493
Washington	0	0	0	1	22	850	0	0	0	1	22	850
Active Sites	18	286	6,873	1	22	850	0	0	0	19	308	7,723
(Marginal Sites)[1]	(0)	(0)	(0)	(0)	(0)	(0)	(0)	(0)	(0)	(0)	(0)	(0)
Percent of U.S.	6	8	11	2	6	10	0	0	0	5	8	9

[1] Marginal sites: Very little or no activity reported during the 2006 calendar year. Marginal sites included in State/Regional totals.

Note: Bcf = Billion cubic feet. MMcf = Million cubic feet. States with no undergrond natural gas storage are shown in *italics*.

Source: Energy Information Administration, GasTran Natural Gas Transportation Information System, Underground Natural Gas Storage Database.

Storage facilities in Washington and Oregon are used primarily to provide seasonal backup to several LDCs located in the northwest and are crucial in maintaining their operational flexibility and system integrity. These storage facilities are also used by some Canadian shipper/customers to support their marketing and operational needs. The import/export facilities of the Northwest Pipeline Company at Sumas, Washington, are used to move natural gas in either direction to storage, depending on marketing conditions.

Overall

Total U.S. working gas capacity and daily deliverability at the beginning of 2007 reached 4.06 Tcf and 85.1 Bcf per day, respectively. Three-hundred and ninety-eight underground natural gas storage facilities were operational in the lower 48 States although 34 were marginal operations that reported little or no activity during 2006. The number of operational underground natural gas storage facilities peaked in 2001 at 418.

In almost all operational aspects, the underground natural gas storage profile of the Midwest Region is larger than that of any of the other five regions. The prevailing cold winters, large population centers, large natural gas pipeline systems, and available geology, have all contributed to major storage development in the region over the past century. The Southwest Region, on the other hand, with its large natural gas production levels and the presence of many large salt-formations, is the second largest source of working gas capacity and daily deliverability in the lower 48 States.

Regional Summary of Underground Natural Gas Storage, by Reservoir Type, 2006

Region	Depleted Gas/Oil Fields			Aquifer Storage			Salt Cavern Storage			Total		
	Sites	Working Gas Capacity (Bcf)	Daily Withdrawal Capability (MMcf)	Sites	Working Gas Capacity (Bcf)	Daily Withdrawal Capability (MMcf)	Sites	Working Gas Capacity (Bcf)	Daily Withdrawal Capability (MMcf)	Sites	Working Gas Capacity (Bcf)	Daily Withdrawal Capability (MMcf)
Central	41	464	4,559	8	91	1,550	1	1	0	50	556	6,109
Midwest	88	917	20,424	31	278	5,855	2	2	85	121	1,197	26,364
Northeast	105	783	14,598	0	0	0	3	6	470	108	789	15,068
Southeast	26	128	3,053	3	7	68	4	38	3,622	33	173	6,743
Southwest	45	898	13,250	1	2	3	21	136	9,799	67	1,036	23,052
Western	18	286	6,873	1	22	850	0	0	0	19	308	7,723
Active Sites	323	3,476	62,757	44	400	8,326	31	183	13,976	398	4,059	85,059
(Marginal Sites)[1]	(31)	(56)	(495)	(1)	(2)	(43)	(2)	(5)	(0)	(34)	(63)	(538)

[1] Marginal sites: Very little or no activity reported during the 2006 calandar year. Marginal sites included in Regional and U.S. totals.

Note: Bcf = Billion cubic feet. MMcf = Million cubic feet. Totals may not sum due to independent rounding.

Source: Energy Information Administration, GasTran Natural Gas Transportation Information System, Underground Natural Gas Storage Database.

Depleted Production Reservoir Underground Natural Gas Storage Well Configuration

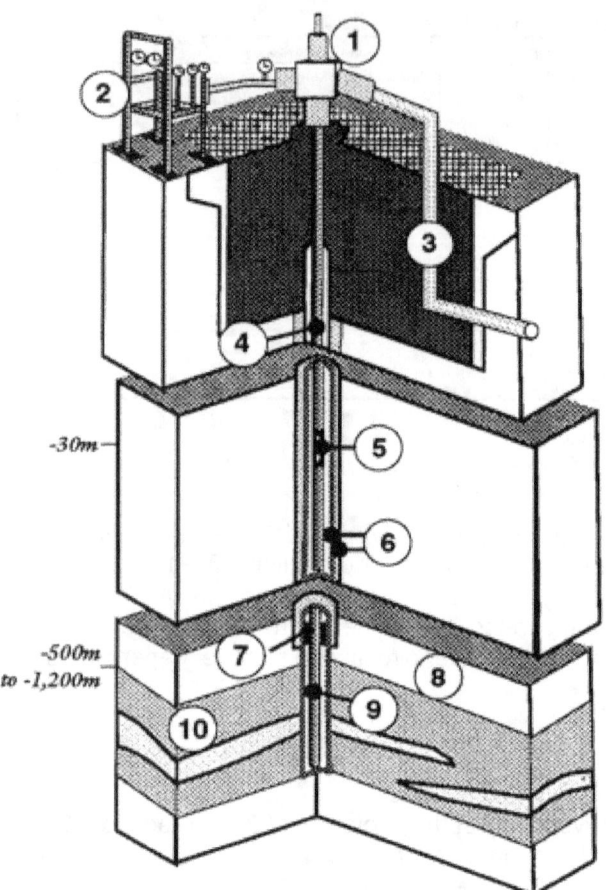

Cross Section of a
Storage Well:

1. Wellhead.
2. Safety Valve Control.
3. Connection Pipe.
4. Flow Tubing String.
5. Automatic Safety Valve.

6. Cemented Casing.
7. Packer.
8. Caprock.
9. Strainers(-500 to -1,200m).
10. Reservoir.

Source: Gaz de France, "Underground Storages Facilities" (June 1992): Recreated by Energy Information Administration, Office of P Management, and Information Services.

Aquifer Underground Natural Gas Storage Reservoir Configuration

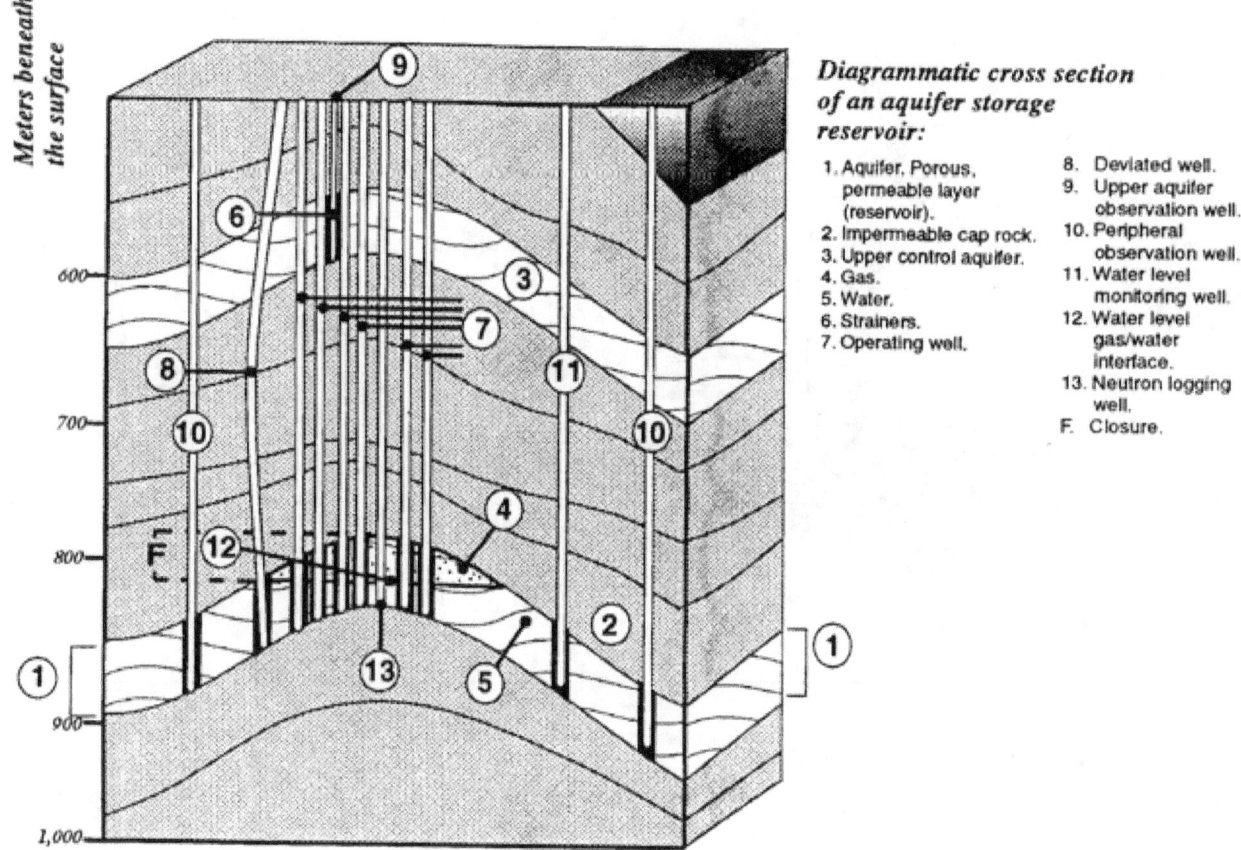

Diagrammatic cross section of an aquifer storage reservoir:

1. Aquifer. Porous, permeable layer (reservoir).
2. Impermeable cap rock.
3. Upper control aquifer.
4. Gas.
5. Water.
6. Strainers.
7. Operating well.
8. Deviated well.
9. Upper aquifer observation well.
10. Peripheral observation well.
11. Water level monitoring well.
12. Water level gas/water interface.
13. Neutron logging well.
F. Closure.

Source: Gaz de France, "Underground Storages Facilities" (June 1992): Recreated by Energy Information Administration, Office of Planning Management, and Information Services.

Salt Cavern Underground Natural Gas Storage Reservoir Configuration

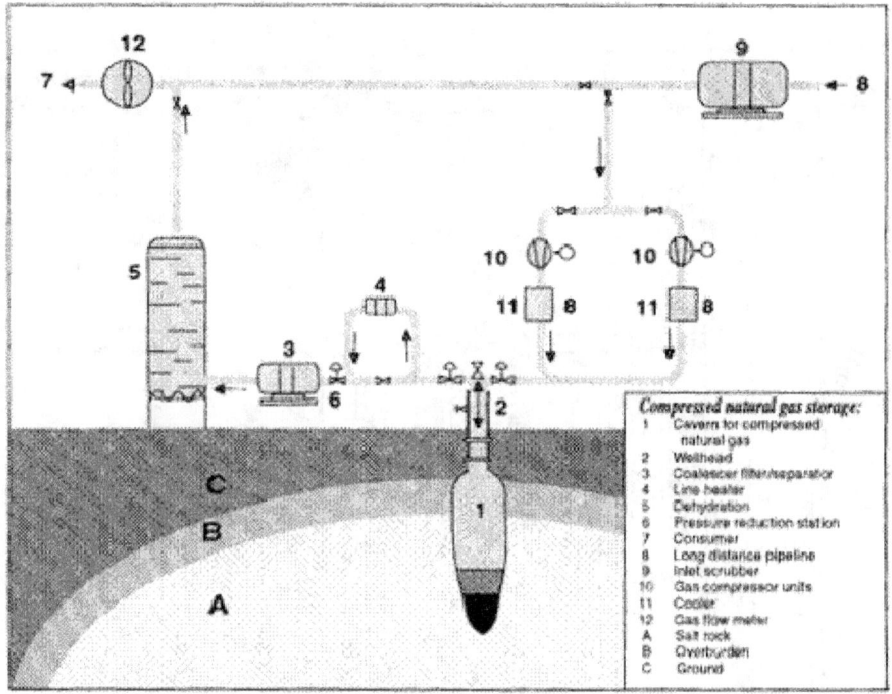

Source: PB-KBB Inc., "Underground Storage and Subsurface Systems"; Recreated by Energy Information Administration, Office of Planning, Management, and Information Services.

Natural Gas Pipeline Development and Expansion

Timing | Determining Market Interest | Expansion Options | Obtaining Approval | Prefiling Process | Approval | Construction | Commissioning

Timing and Steps for a New Project

An interstate natural gas pipeline construction or expansion project takes an average of about three years from the time it is first announced until the new pipe is placed in service. The project can take longer if it encounters major environmental obstacles or public opposition.

A pipeline development or expansion project involves several steps:

- Determining demand/market interest
- Publicly announcing the project
- Obtaining regulatory approval
- Construction and testing

Determining Market Interest and Public Announcement

To gauge the level of market interest, an open season is held for 1-2 months, giving potential customers an opportunity to enter into a nonbinding agreement to sign up for a portion of the capacity rights that will be available. If enough interest is shown during the open season, the sponsors will develop a preliminary project design and move forward. If not enough interest is evident, the project will most likely be dropped or placed on indefinite hold.

Expansion and Development Options

Options for creating additional pipeline capacity include:

- Building an entirely new pipeline
- Converting an oil or product pipeline to a natural gas pipeline
- Adding a parallel pipeline along a segment of pipeline, called looping
- Installing a lateral or extension off the existing mainline
- Upgrading and expanding facilities, such as compressor stations, along an existing route. This option is usually the quickest, least expensive, and has the least environmental impacts.

Development and Expansion Process

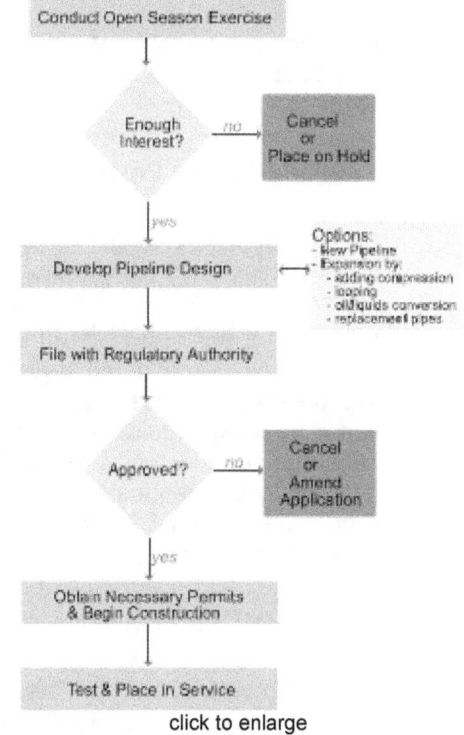

click to enlarge

More sources of information related to pipeline development & expansion ...

Natural Gas Pipeline Expansions in 2005 - report
Natural Gas Pipeline Expansions in 2004 - report
Natural Gas Pipeline & Storage Expansions in 2003 - report
Natural Gas Pipeline Expansions in 2002 - report
Status of Pipeline System Capacity 2000-01 - report
Natural Gas Pipeline Expansions in 2000 - report
Deliverability on the Interstate System - 1998 - report

Obtaining Construction Approval

Developing the final project design and obtaining first financial commitments from potential customers may take from three to six months. Then, the project specifications are filed with the appropriate regulatory agency.

Pre-filing Review Process

Of the proposed project involves an interstate pipeline, that is, it falls under the jurisdiction of the Federal Energy Regulatory Commission (FERC), the project sponsor has the option of either requesting that a National Environmental Policy Act (NEPA) pre-filing review be initiated during the early states of project design, or waiting until later and filing with FERC under the traditional application review process.

The pre-filing process is designed to facilitate and expedite the review of natural gas pipeline projects that would normally require FERC to prepare an environmental assessment, an environmental impact statement, or a historic preservation review as part of the traditional review process. The project sponsor must notify and request that the various regulatory agencies be involved in evaluating the project if a pre-filing review from NEPA is filed. In this case, FERC staff will take the lead in scheduling and coordinating the approval steps.

Approval of the Regulating Authority

A FERC review of an interstate pipeline project takes from 5-18 months, with an average time of 15 months. No data are available on the average time for obtaining approval from an individual State agency. Usually, approval by the regulating authority is conditional, but most often the conditions do not constitute a significant impediment. The project sponsor must then either accept or reject the conditions or reapply with an alternative plan.

Construction

Pipeline construction is usually completed within 18 months and sometimes in as little as 6 months. Construction can be delayed because additional time may be needed to acquire local permits from towns and land-use agencies located along the proposed construction route.

Commissioning and Testing

Commissioning and testing the completed pipeline project usually takes about one to three weeks. This process involves subjecting the new segments of the pipeline to hydrostatic testing (water fill under high pressure) or other tests of the line in-place. Line packing, which involves filling the line with the initial baseload volume of natural gas, is usually needed only on a new pipeline or on larger expansion projects.

Natural Gas Import/Export Pipelines

Currently the United States has 55 locations where natural gas can be exported or imported.

- 24 locations are for imports only
- 19 locations are for exports only
- 12 locations are for both imports and exports
- 5 locations are liquefied natural gas (LNG) import facilities

Imported natural gas currently (2006) represents almost 17 percent of the gas consumed in the United States annually, compared with 11 percent just 10 years ago.

Forty natural gas pipelines, representing approximately 23 billion cubic feet (Bcf) per day of capacity, import and export natural gas between the United States and Canada or Mexico.

U.S. Natural Gas Import & Export Locations

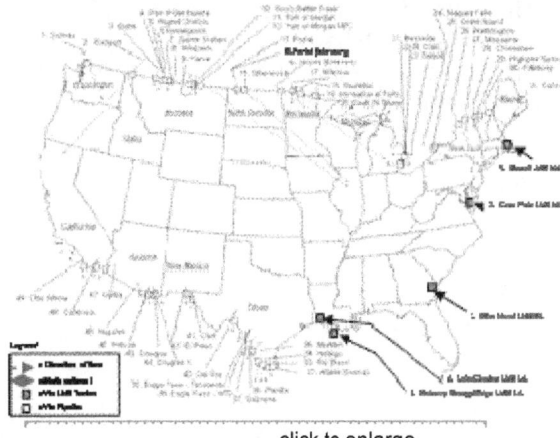

click to enlarge

More information related to imports and exports…

U.S./Canada/Mexico Natural Gas Import/Export Points List - table

Between 1990 and 2006, import pipeline capacity from Canada increased by 169 percent (to 17.3 Bcf per day) and from Mexico by 147 percent (to 0.9 Bcf per day). During the same period, export capacity to Canada more than tripled (to 4.2 Bcf per day) while export capacity to Mexico quadrupled (to 3.6 Bcf per day).

In 2006, the United States received 99.8 percent of its pipeline- imported natural gas from Canada with the remainder from Mexico. Canada also accounted for 60 percent of pipeline natural gas exports, and Mexico, 40 percent.

In 2005, the top five import points accounted for about 70 percent of all natural gas brought into the United States via pipeline. They are:

- Port of Morgan, Montana (Northern Border Pipeline)
- Eastport, Idaho (Gas Transmission Northwest)
- Sherwood, North Dakota (Alliance Pipeline Company)
- Noyes, Minnesota (Great Lakes Gas Transmission Company)
- Noyes, Minnesota (Viking Gas Transmission Company)

Five relatively small border crossing points (four in Montana and one in North Dakota) were installed during the past decade, primarily to facilitate the movement of local production gas to processing plants or to pipeline receipt points located on the opposite side of the border. The relatively small level of natural gas flowing through these points is not counted as imports to the respective receiving country.

U.S. natural gas import and export activities are regulated under Section 3 of the Natural Gas Act of 1938 by the U.S. Department of Energy and the Federal Energy Regulatory Commission (FERC).   While FERC is responsible for review and approval of the actual siting, construction, and operation of natural gas import and export facilities, DOE is responsible for authorization of the contracts governing the importing and exporting of natural gas.

Currently, there are 55 locations at which natural gas can be exported or imported into the United States, including 6 LNG (liquefied natural gas) facilities in the continental United States and Alaska (There is a seventh U.S. LNG import facility located in Puerto Rico). At 24 of these locations natural gas or LNG currently can only be imported; while at 19 they may only be exported (1 LNG export facility is located in Alaska). At 12 of the 55 locations natural gas may, and sometimes does, flow in both directions, although at each of these sites the flow is primarily either import or export.

Appendix A. Combined 'Natural Gas Transportation' Maps

U.S. Natural Gas Pipeline Network

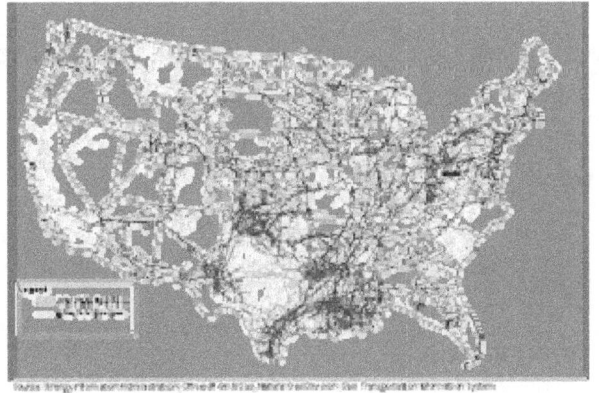

Major Natural Gas Supply Basins Relative to Natural Gas Pipeline Transportation Corridors

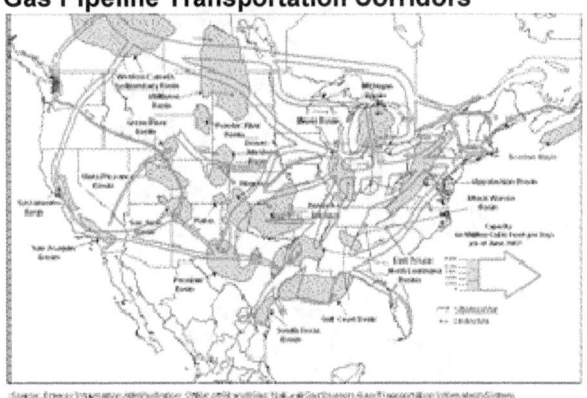

U.S. Regional Breakdown

States (in Grey) Highly Dependent on Interstate Pipelines for Natural Gas Supplies

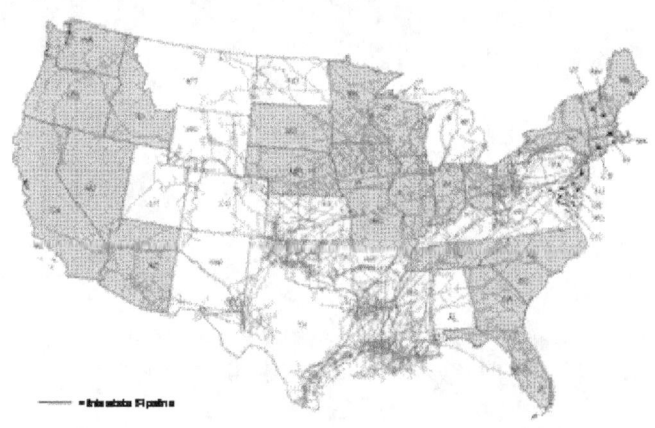

U.S. Underground Natural Gas Storage Facilities

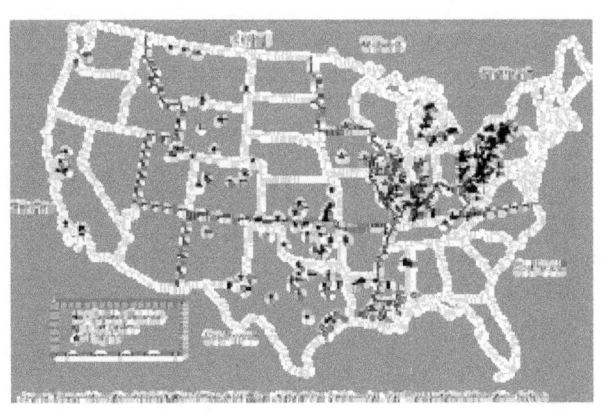

U.S. Natural Gas Pipeline Import and Export Locations

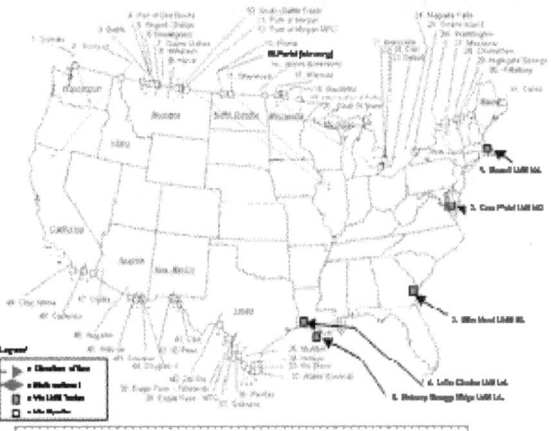

Major U.S. Natural Gas Transportation Corridors

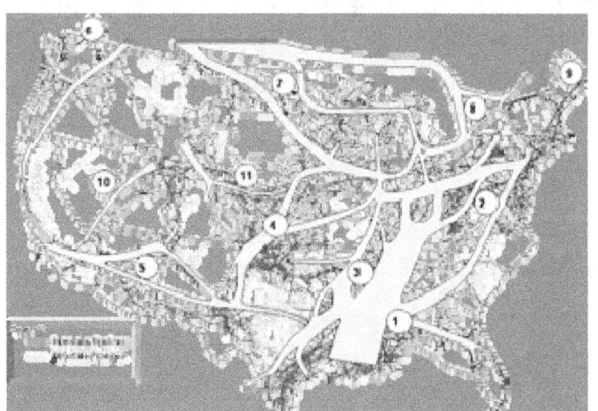

U.S. Natural Gas Pipeline Compressor Stations

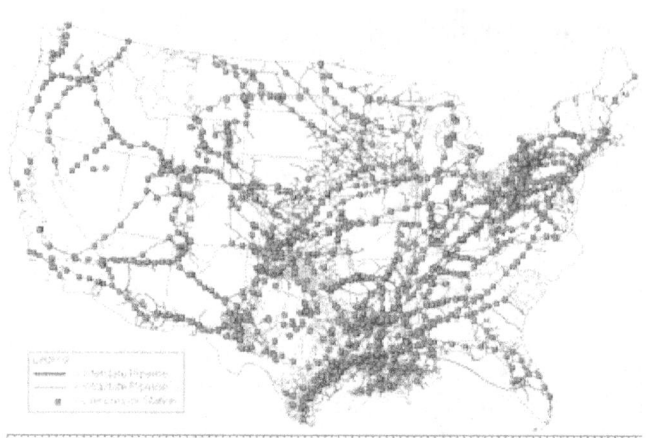

Major North American Natural Gas Market Centers & Hubs

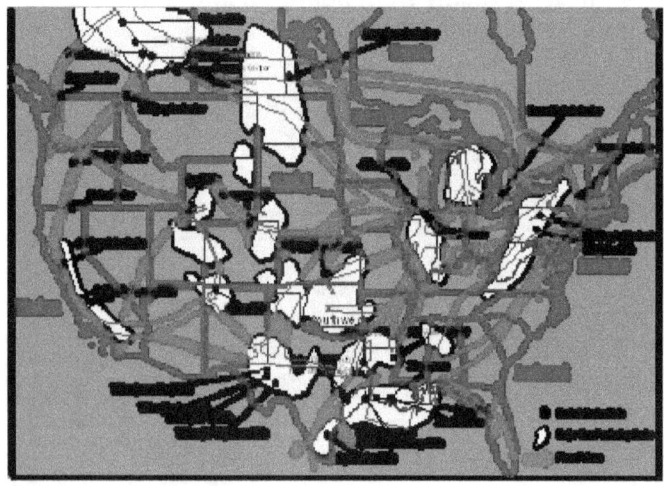

Interregional Natural Gas Transmission Pipeline Capacity

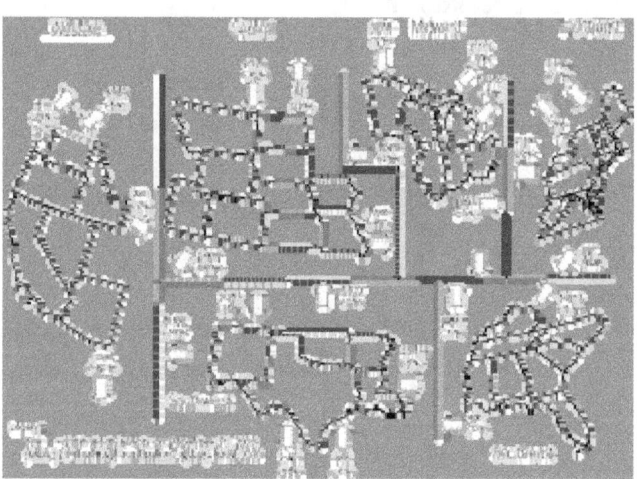

The EIA has determined that the informational map displays here do not raise security concerns, based on the application of the Federal Geographic Data Committee's *Guidelines for Providing Appropriate Access to Geospatial Data in Response to Security Concerns*.

Thirty Largest U.S. Interstate Natural Gas Pipeline Systems, 2005

(Ranked by system capacity, million cubic feet per day (MMcf/d))

Pipeline Name	Market Regions Served	Primary Supply Regions	States in Which Pipeline Operates	Transported (billion cubic dekatherms)	System Capacity (MMcf/d) [1]	System Mileage
Columbia Gas Transmission Co.	Northeast	Southwest, Appalachia	DE, PA, MD, KY, NC, NJ, NY, OH, VA, WV	3,431	8,700	10,354
Transcontinental Gas Pipeline Co.	Northeast, Southeast	Southwest	AL, GA, LA, MD, MS, NC, NY, SC, TX, VA, GM	3,338	8,161	10,469
Northern Natural Gas Co.	Central, Midwest	Southwest	IA, IL, KS, NE, NM, OK, SD, TX, WI, GM	1,195	7,923	15,854
ANR Pipeline Co.	Midwest	Southwest	AR, IA, IL, IN, KS, KY, LA, MI, MO, MS, NE, OH, OK, WI, GM	2,815	6,844	9,616
Tennessee Gas Pipeline Co.	Northeast, Midwest	Southwest, Canada	AR, KY, LA, MA, NY, OH, PA, TN, TX, WV, GM	1,920	6,686	13,302
Texas Eastern Transmission Corp.	Northeast	Southwest	AL, AR, IL, IN, KS, KY, LA, MI, MO, MS, NJ, NY, OH, OK, PA, TX, WV, GM	1,364	6,523	9,179
El Paso Natural Gas Co.	Western, Southwest	Southwest	AZ, CO, NM, TX	4,864	6,152	10,661

Company	Region	Source	States	Col1	Col2	Col3
Dominion Transmission Co.	Northeast	Southwest, Appalachia	PA, MD, NY, OH, VA, WV	1,344	5,734	3,142
Northwest Pipeline Corp.	Western	Canada, Central	CO, ID, OR, UT, WA, WY	700	4,500	4,046
Natural Gas Pipeline Co. of America	Midwest	Southwest	AR, IA, IL, KS, LA, MO, NE, OK, TX, GM	2,69	4,485	9,111
Southern Natural Gas Co.	Southeast	Southwest	AL, GA, LA, MS, SC, TN, TX, GM	937	3,365	7,671
Centerpoint Gas Transmission Co.	Southwest	Southwest	AR, KS, LA, OK, TX	928	3,339	6,182
Gulf South Pipeline Co.	Southeast, Southwest	Southwest	AL, FL, LA, MS, TX, GM	1,015	3,038	6,580
Colorado Interstate Gas Co.	Central	Central, Southwest	CO, KS, OK, TX, WY	939	3,000	3,996
Texas Gas Transmission Corp.	Midwest	Southwest	AR, IN, KY, LA, MS, OH, TN	2,178	2,979	5,643
Great Lakes Gas Transmission Co.	Midwest	Canada	MI, MN, WI	958	2,859	2,115
Panhandle Eastern Pipeline Co.	Midwest	Southwest	IL, IN, KS, MI, MO, OH, OK, TX	709	2,840	6,445
Gas Transmission Northwest Corp.	Western	Canada	ID, OR, WA	767	2,636	1,356
Northern Border Pipeline Co.	Midwest, Central	Canada	IA, IL, IN, MN, MT, ND, SD	898	2,496	1,399
Southern Star Central Pipeline Co.	Central	Central	CO, KS, MO, NE, OK, TX, WY	354	2,451	5,788
National Fuel Gas Supply Co.	Northeast	Canada, Appalachia	NY, PA	417	2,312	1,504
Questar Pipeline Co.	Central	Central	CO, UT, WY	379	2,192	1,745
Florida Gas Transmission Co.	Southeast	Southwest	AL, FL, LA, MS, TX, GM	757	2,190	4,867

Algonquin Gas Transmission Co.	Northeast	Southwest	CT, MA, NJ, NY, RI	346	2,174	1,103
Columbia Gulf Transmission Co.	Southeast, Northeast	Southwest	KY, LA, MS, TN, GM	2,041	2,156	4,105
Alliance Pipeline Co. (US)	Midwest	Canada	ND, MN, IA, IL	652	2,053	888
Wyoming Interstate Gas Co.	Central	Central	CO, WY	594	1,997	585
Kern River Gas Transmission Co.	Western	Central	CA, NV, UT, WY	718	1,833	1,680
High Island Offshore System	Southwest	Gulf of Mexico	LA, GM	234	1,800	212
Trunkline Gas Co.	Midwest	Southwest	AR, IL, IN, KY, LA, MS, OH, TN, TX	606	1,680	3,558
Sub-total				37,398	115,098	163,156
Other Interstate Systems (79)	- -	- -	- -	10,242	33,235	49,531
Total				47,640	148,333	212,687

¹ Capacity levels are reported to FERC in Btu, dekatherms, or volumetric units. For this presentation, reported capacity figures are presented as volumetric (MMcf/d = million cubic feet per day) assuming a conversion factor of 1 MMcf/d = 1 MDth/d (thousand dekatherms per day) = 1 Bbtu/d (billion btus per day).

Note: GM = Gulf of Mexico.

Source: Federal Energy Regulatory Commission (FERC), **Mileage & Transport**: FERC Form 2 & 2A "Major and Non-major Natural Gas Pipeline Annual Report," **Capacity**: FERC Annual Peak Day Capacity Report Section 284.13(d).

Estimated Natural Gas Pipeline Mileage in the Lower 48 States, 2006

Region/State	Pipeline Mileage	Region/State	Pipeline Mileage	Region/State	Pipeline Mileage	Region/State	Pipeline Mileage	Region/State	Pipeline Mileage	Region/State	Pip Mil
Central		**Midwest**		**Northeast**		**Southeast**		**Southwest**		**Western**	
Colorado	7,405	Illinois	11,908	Connecticut	619	Alabama	4,691	Arkansas	6,201	Arizona	5
Iowa	5,347	Indiana	4,704	Delaware	265	Florida	4,746	Louisiana	18,358	California	11
Kansas	15,284	Michigan	9,706	Maine	607	Georgia	3,360	New Mexico	6,728	Idaho	1
Missouri	3,771	Minnesota	4,434	Maryland/DC	972	Kentucky	6,824	Oklahoma	18,494	Nevada	1
Montana	3,861	Ohio	7,656	Massachusetts	939	Mississippi	9,484	Texas	57,160	Oregon	1
Nebraska	5,346	Wisconsin	3,320	New Hampshire	291	North Carolina	2,484		106,941	Washington	2
North Dakota	1,873		41,728	New Jersey	1,516	South Carolina	2,265				24
South Dakota	1,242			New York	4,741	Tennessee	4,273				
Utah	3,034			Pennsylvania	8,546		38,127	Gulf Mexico[1]	9,329		
Wyoming	7,442			Rhode Island	100						
	54,605			Vermont	53						
				Virginia	2,493			**Total US Pipeline Mileage**			300
				West Virginia	3,739			Total Interstate[1]			213
					24,881			Total Non-interstate [2]			86

[1] In the Gulf of Mexico some large-scale gathering systems are FERC jurisdictional and are therefore counted as interstate.

[2] Includes intrastate transmission and non-FERC jurisdictional large diameter gathering systems or headers. Local distribution company (LDC) mileage excluded.

Note: All mileage is approximate. Includes looped pipeline segments. Approximately 72 percent of Interstate pipeline systems are made up of pipeline diameters excee... 16 inches while only 35 percent of non-interstate pipeline systems are 16 inches or larger.

Source: Energy Information Administration, Gas Transportation Information System, Pipeline Map Files.

Alphabetical List of Pipeline Companies: Links to U.S. Natural Gas Pipeline Information

The links below will either direct the user to a narrative describing the system, a pipeline system map, a FERC prescribed "Informational Postings" page, or a FERC Tariff Sheet.

Pipeline Name	Type of System	Regions of Operations
Acadian Gas Pipeline System	Intrastate	Southwest
Algonquin Gas Transmission Co	Interstate	Northeast
Alliance Pipeline Co	Interstate	Central, Midwest
Anaconda Pipeline System	Gathering	Gulf of Mexico
ANR Storage Co	Interstate	Midwest
Arkansas Oklahoma Gas Co	Intrastate	Southwest
Arkansas Western Pipeline Co	Intrastate	Southwest
Atmos Energy Pipeline – Texas	Intrastate	Southwest
Atmos Pipeline & Storage Co (MS)	Intrastate	Southeast
Bighorn Gas Gathering Header (WY)	Gathering	Central
Black Marlin Offshore Pipeline	Interstate	Gulf of Mexico
Blue Dolphin Pipeline Co	Gathering	Gulf of Mexico
Bluewater System	Gathering	Gulf of Mexico
B-R (USG) Pipeline Co	Interstate	Southeast
Bridgeline Gas Systems	Intrastate	Southwest
California Gas Transmission Co	Intrastate	Western
Canyon Chief Pipeline System	Gathering	Gulf of Mexico
Cardinal Pipeline Co (NC)	Intrastate	Southeast
Carolina Gas Transmission Co	Interstate	Southeast
Cascade Natural Gas Co	Intrastate	Western
CCNG Transmission System	Intrastate	Southwest
Centerpoint Energy Pipeline Co	Interstate	Central, Southwest
Centerpoint Mississippi River Trans Co	Interstate	Central, Southwest, Midwest
Chandeleur Pipeline Co	Interstate	Southeast, Gulf of Mexico
Cleopatra Gathering System	Gathering	Gulf of Mexico
Colorado Interstate Gas Co	Interstate	Central, Southwest
Columbia Gas Transmission Co	Interstate	Northeast, Midwest, Southeast
Columbia Gulf Transmission Co	Interstate	Southwest, Southeast, Gulf of Mexico
Consumers Gas Co (MI)	Intrastate	Midwest
Crossroads Pipeline Co	Interstate	Midwest
Cypress Pipeline Co	Intrastate	Southwest
Dauphin Island Gathering System	Gathering	Gulf of Mexico, Southeast
Destin Pipeline LP	Gathering	Gulf of Mexico, Southeast
Discovery Pipeline Co	Gathering	Gulf of Mexico, Southwest
Dominion Cove Point LNG LP	Interstate	Northeast
Dominion East Ohio (OH)	Intrastate	Midwest
Dominion Gas Transmission Co	Interstate	Midwest
Dominion Hope Gas Co (WV)	Intrastate	Northeast

Dominion Transmission Corp	Interstate	Northeast
East Breaks Gathering System	Gathering	Gulf of Mexico
East Tennessee Natural Gas Co	Interstate	Southeast, Northeast
Eastern Shore Natural Gas Co	Interstate	Northeast
El Paso Natural Gas Co	Interstate	Western, Southwest
Empire Gas Pipeline Co (NY)	Intrastate	Northeast
Enbridge Pipelines (Alabama Intra)	Intrastate	Southeast
Enbridge Pipelines (AlaTenn)	Interstate	Southeast
Enbridge Pipelines (East Texas)	Intrastate	Southwest
Enbridge Pipelines (KPC) (KS,MO)	Intrastate	Central
Enbridge Pipelines (Louisiana Intra)	Intrastate	Southwest
Enbridge Pipelines (MidLa)	Interstate	Southwest, Southeast
Enbridge Pipelines (North Texas)	Intrastate	Southwest
Enbridge Pipelines (UTOS)	Interstate	Gulf of Mexico, Southwest
Energy Transfer East Texas Pipeline Co	Intrastate	Southwest
Enogex Pipeline System	Intrastate	Southwest
EPGT Texas Intrastate Pipeline	Intrastate	Southwest
Equitrans Inc	Interstate	Northeast
Evangeline Gas Pipeline Co	Intrastate	Southwest
Falcon Gas Pipeline System	Gathering	Gulf of Mexico
Ferndale Pipeline Co	Intrastate	Western
Florida Gas Transmission Co	Interstate	Southeast, Southwest
Fort Union Gathering Header (WY)	Gathering	Central
Garden Banks Gas Pipeline System	Gathering	Gulf of Mexico
Gas Transmission Northwest	Interstate	Western
Granite State Gas Transportation Co	Interstate	Northeast
Great Lakes Gas Transmission Ltd	Interstate	Midwest
Green Canyon System	Gathering	Gulf of Mexico
Guadalupe Pipeline Co	Intrastate	Southwest
Guardian Pipeline Co	Interstate	Midwest
Gulf Coast Pipeline System	Intrastate	Southwest
Gulf South Pipeline Co	Interstate	Southwest, Southeast
Gulfstream Natural Gas System	Interstate	Southeast
High Island Offshore System	Gathering	Gulf of Mexico, Southwest
Horizon Pipeline Co	Interstate	Midwest
Houston Gas Pipeline Co	Intrastate	Southwest
Iroquois Gas Transmission Co	Interstate	Northeast
Jonah Gas Gathering Co (WY)	Gathering	Central
Kelso-Beaver Pipeline Co	Interstate	Western
Kern River Transmission Co	Interstate	Western
Kinder-Morgan Border Pipeline	Intrastate	Southwest
Kinder-Morgan North Texas Pipeline	Intrastate	Southwest
Kinder-Morgan South Texas Pipeline	Intrastate	Southwest
Kinder-Morgan Tejas Gas Pipeline	Intrastate	Southwest
Kinder-Morgan Texas Pipeline Co	Intrastate	Southwest

KM Interstate Pipeline Co	Interstate	Central, Southwest
KO Gas Transmission Co	Interstate	Midwest
Lost Creek Gathering Co (WY)	Gathering	Central
Louisiana Intrastate Gas (LIG) Co	Intrastate	Southwest
Magnolia Gathering Lateral	Gathering	Gulf of Mexico
Manta Ray Gathering System	Gathering	Gulf of Mexico
Maritimes & Northeast Pipeline Co	Interstate	Northeast
MarkWest Intrastate Pipeline Co	Intrastate	Southwest
Matagorda Offshore Pipeline System	Gathering	Gulf of Mexico
MichCon Gas Co	Interstate	Midwest
Midwestern Gas Transmission Co	Interstate	Midwest, Southeast
MIGC Pipeline Co	Interstate	Central
Mississippi Canyon Gathering System	Gathering	Gulf of Mexico
Missouri Interstate Gas Pipeline LLC	Intrastate	Midwest
Mojave Pipeline Co	Interstate	Western
National Fuel Gas Distribution Co (NY)	Intrastate	Northeast
National Fuel Gas Supply Corp	Interstate	Northeast
Natural Gas PL Co of America	Interstate	Southwest, Central, Midwest, Gulf of Mexico
Nautilus Pipeline Co	Interstate	Gulf of Mexico, Southwest
Nemo Pipeline Co	Gathering	Gulf of Mexico
NORA Gas Transmission Co	Interstate	Midwest
NorNew/Norse Pipeline System (NY)	Intrastate	Northeast
North Baja Pipeline Co	Interstate	Western
NorthCoast Gas Transmission Co (OH)	Intrastate	Midwest
Northern Border Pipeline Co	Interstate	Central, Midwest
Northern Illinois Gas Co (NICOR) (IL)	Intrastate	Midwest
Northern Indiana Public Service Co	Intrastate	Midwest
Northern Natural Gas Co	Interstate	Southwest, Central, Midwest, Gulf of Mexico
Northern Utilities Inc (ME)	Intrastate	Northeast
Northwest Natural Gas Co	Intrastate	Western
Northwest Pipeline Co	Interstate	Western, Central
NorthWestern Energy (MT)	Intrastate	Central
Oasis Gas Pipeline Co	Intrastate	Southwest
Okeanos Deepwater System	Gathering	Gulf of Mexico
Oklahoma Natural Gas Co	Intrastate	Southwest
OkTex Pipeline Co	Interstate	Southwest
Overland Trail Transmission Co (WY)	Intrastate	Central
(Questar) Overthrust Pipeline Co	Interstate	Central
Ozark Gas Transmission Co	Interstate	Southwest
Panhandle Eastern PL Co	Interstate	Southwest, Central, Midwest, Gulf of Mexico
Pauite Pipeline Co (NV)	Intrastate	Western
Phoenix Gathering System	Gathering	Gulf of Mexico
Portland Natural Gas Transportation System	Interstate	Northeast
Pub Svc Co of North Carolina (NC)	Intrastate	Southeast
Public Service Co of New Mexico	Intrastate	Southwest

Questar Gas Co (UT, WY)	Intrastate	Central
Questar Pipeline Co	Interstate	Central
Regency Intrastate Gas Co	Intrastate	Southwest
Rocky Mountain Natural Gas Co (CO)	Intrastate	Central
Sabine Pipeline Co	Interstate	Southwest
Saginaw Bay Pipeline (MI)	Intrastate	Midwest
San Diego Gas & Electric Co (CA)	Intrastate	Western
Sandhill Pipeline Co (NC)	Intrastate	Southeast
Sea Robin Pipeline System	Interstate	Gulf of Mexico, Southwest
Southern California Gas Co (CA)	Intrastate	Western
Southern Natural Gas Co	Interstate	Southwest, Southeast, Gulf of Mexico
Southern Star Central Pipeline Co	Interstate	Central, Southwest
Southern Trails Pipeline (Questar)	Interstate	Western
Southwest Gas Co (CA, NV)	Intrastate	Western
SouthWestern Energy Pipeline Co	Intrastate	Southwest
Southern Union Intrastate Pipelines	Intrastate	Southwest
St. Lawrence Gas Co	Interstate	Northeast
Stingray Pipeline System	Interstate	Gulf of Mexico, Southwest
Sumas International Pipeline	Intrastate	Western
Tengasco Pipeline Co (TN)	Intrastate	Southeast
Tennessee Gas Pipeline Co	Interstate	Southwest, Southeast, Midwest, Northeast
Texas Eastern Transmission Co	Interstate	Southwest, Southeast, Midwest, Northeast
Texas Gas Transmission Co	Interstate	Southwest, Southeast, Midwest
Texas Intrastate Pipeline Co	Intrastate	Southwest
Tidelands Pipeline System	Intrastate	Southwest
Trailblazer Pipeline Co	Interstate	Central
Trans-Union Pipeline Co	Interstate	Southwest
TransColorado Gas Transmission Co	Interstate	Central, Southwest
Transcontinental Gas Pipeline Co	Interstate	Southwest, Southeast, Northeast, Gulf of Mexico
Transwestern Gas Pipeline Co	Interstate	Southwest, Western
Triton Gathering Lateral	Gathering	Gulf of Mexico
Trunkline Gas Co	Interstate	Southwest, Southeast, Midwest, Gulf of Mexico
Tuscarora Gas Transmission Co	Interstate	Western
Typhoon Gas Gathering System	Gathering	Gulf of Mexico
Vanderbuilt Pipeline System	Intrastate	Southwest
Vector Pipeline Co	Interstate	Midwest
Venice Gas Gathering System	Gathering	Gulf of Mexico
Vermont Gas Systems Inc	Interstate	Northeast
Viking Gas Transmission Co	Interstate	Midwest
Viosca Knoll Gathering System	Gathering	Gulf of Mexico
Virginia Natural Gas Co	Intrastate	Northeast
West Texas Gas Co	Interstate	Southwest
Williston Basin Interstate PL Co	Interstate	Central
Wyoming Interstate Co	Interstate	Central

Regional Underground Natural Gas Storage, Beginning of 2007

Region/ State	Depleted-Reservoir Storage			Aquifer Storage			Salt-Cavern Storage			Total		
	Sites	Working Gas Capacity (Bcf)	Daily Withdrawal Capability (MMcf)	Sites	Working Gas Capacity (Bcf)	Daily Withdrawal Capability (MMcf)	Sites	Working Gas Capacity (Bcf)	Daily Withdrawal Capability (MMcf)	Sites	Working Gas Capacity (Bcf)	Daily Withdrawal Capability (MMcf)
Central Region												
Colorado	9	42	1,088	0	0	0	0	0	0	9	42	1,088
Iowa	0	0	0	4	75	1,025	0	0	0	4	75	1,025
Kansas	18	117	2,348	0	0	0	1	1	0	19	118	2,348
Missouri	0	0	0	1	11	350	0	0	0	1	11	350
Montana	5	196	300	0	0	0	0	0	0	5	196	300
Nebraska	1	16	169	0	0	0	0	0	0	1	16	169
North Dakota	0	0	0	0	0	0	0	0	0	0	0	0
South Dakota	0	0	0	0	0	0	0	0	0	0	0	0
Utah	1	51	427	2	1	100	0	0	0	3	52	527
Wyoming	7	42	227	1	4	75	0	0	0	8	46	302
Total Sites	41	464	4,559	8	91	1,550	1	1	0	50	556	6,109
(Marginal Sites)[1]	(7)	(4)	(145)	(0)	(0)	(0)	(1)	(1)	(0)	(8)	(5)	(145)
Midwest Region												
Illinois	11	51	835	18	256	5,294	0	0	0	29	307	6,129
Indiana	10	14	261	12	20	501	0	0	0	22	34	762
Michigan	43	632	14,636	0	0	0	2	2	85	45	634	14,721
Minnesota	0	0	0	1	2	60	0	0	0	1	2	60
Ohio	24	220	4,692	0	0	0	0	0	0	24	220	4,692
Total Sites	88	917	20,424	31	278	5,855	2	2	85	121	1,197	26,364
(Marginal Sites)[1]	(8)	(8)	(119)	(1)	(2)	(43)	(0)	(0)	(0)	(9)	(10)	(162)
Northeast Region												
Maryland	1	17	400	0	0	0	0	0	0	1	17	400
New York	23	116	1,840	0	0	0	1	2	145	24	118	1,985
Pennsylvania	49	405	8,635	0	0	0	0	0	0	49	405	8,6358
Virginia	1	1	22	0	0	0	2	4	325	3	5	347
West Virginia	31	244	3,701	0	0	0	0	0	0	31	244	3,701
Total Sites	105	783	14,598	0	0	0	3	6	470	108	789	15,068
(Marginal Sites)[1]	(6)	(29)	(32)	(0)	(0)	(0)	(0)	(0)	(0)	(6)	(29)	(32)
Southeast Region												
Alabama	1	8	210	0	0	0	1	7	600	2	15	810
Kentucky	20	80	1,753	3	7	68	0	0	0	23	87	1,821
Mississippi	4	39	1,070	0	0	0	3	31	3,022	7	70	4,092
Tennessee	1	1	20	0	0	0	0	0	0	1	1	20
Total Sites	26	128	3,053	3	7	68	4	38	3,622	33	173	6,743
(Marginal Sites)[1]	(4)	(2)	(22)	(0)	(0)	(0)	(0)	(0)	(0)	(4)	(2)	(22)

Southwest Region

Arkansas	2	15	231	0	0	0	0	0	0	2	15	231
Louisiana	8	286	3,999	0	0	0	6	47	3,153	14	333	7,152
New Mexico	2	54	310	1	2	3	0	0	0	3	56	313
Oklahoma	13	188	3,870	0	0	0	0	0	0	13	188	3,870
Texas	20	355	4,840	0	0	0	15	89	6,646	35	443	11,486
Total Sites	45	898	13,250	1	2	3	21	136	9,799	67	1,036	23,052
(Marginal Sites)[1]	(6)	(13)	(177)	(0)	(0)	(0)	(1)	(4)	(0)	(7)	(17)	(177)

Western Region

California	12	272	6,380	0	0	0	0	0	0	12	272	6,380
Oregon	6	14	493	0	0	0	0	0	0	6	14	493
Washington	0	0	0	1	22	850	0	0	0	1	22	850
Total Sites	18	286	6,873	1	22	850	0	0	0	19	3087	7,723
(Marginal Sites)[1]	(0)	(0)	(0)	(0)	(0)	(0)	(0)	(0)	(0)	(0)	(0)	(0)

Total U.S. Sites	323	3,476	62,757	44	400	8,326	31	183	13,976	398	4,059	85,059
(Marginal Sites)[1]	(31)	(56)	495)	(1)	(2)	(43)	(2)	(5)	(0)	(34)	(63)	(538)

[1]Marginal sites: very little or no activity reported during the 2006 calendar year. However, these sites are included in regional and State summary totals.

Note: Bcf = Billion cubic feet. MMcf = Million cubic feet. Totals may not sum due to independent rounding.

Source: Energy Information Administration, GasTran Natural Gas Transportation Information System, Underground Natural Gas Storage Database.

www.ingramcontent.com/pod-product-compliance
Lightning Source LLC
Chambersburg PA
CBHW080559180526
45168CB00007B/2717